Hands-on Pattern Mining

Uday Kiran Rage

Hands-on Pattern Mining

Theory and Examples with PAMI, Sklearn, Keras, and TensorFlow

Springer

Uday Kiran Rage
Division of Information Systems
University of Aizu
Aizu-Wakamatsu, Fukushima, Japan

ISBN 978-981-96-6793-2 ISBN 978-981-96-6791-8 (eBook)
https://doi.org/10.1007/978-981-96-6791-8

This Springer imprint is published by the registered company Springer Nature Singapore Pte Ltd.
The registered company address is: 152 Beach Road, #21-01/04 Gateway East, Singapore 189721, Singapore

To my beloved wife,

You are the cornerstone of our family, holding everything together with grace, love, and unwavering dedication. Every day, you give so much of yourself to care for our children, nurture our home, and ensure our family thrives. The love and warmth you infuse into every detail of our lives do not go unnoticed. They are the quiet miracles that make our world a better place.

From sleepless nights to laughter-filled days, from small, unnoticed tasks to big moments that define our lives, you are there, giving selflessly and loving unconditionally. You have made our house a home and turned our family into a haven of happiness and love.

This book reflects my gratitude for everything you do—the countless hours, the endless energy, and the boundless love you pour into our children, family, and me. You are a remarkable mother, a devoted wife, and the true heart of our family. Without you, none of this would be possible.

With all my love and deep appreciation, this book is dedicated to you, the love of my life and the guiding light of our family.

Preface

Pattern mining, a fundamental concept in data science and machine learning, is at the heart of discovering valuable insights from real-world big data. This book delves into the intricate world of pattern mining, offering a comprehensive theoretical and practical guide for beginners and seasoned practitioners.

In the current data-driven era, where information overload is a significant challenge for enterprises, the ability to uncover meaningful patterns from big data has become indispensable. Pattern mining enables us to make informed decisions by discovering useful information in numerous forms (e.g., frequent, recurring, high-utility, and periodic patterns) across various domains (e.g., retail, healthcare, and finance).

This book is designed to be a practical companion, blending theoretical foundations with hands-on techniques and applications. It covers a spectrum of topics ranging from basic concepts to advanced techniques. The practical examples in this book are covered using an open-source PAttern MIning (PAMI) library. The implementation code and the sample datasets accompanying the examples in this book can be accessed on our GitHub repository: https://github.com/UdayLab/Hands-on-Pattern-Mining.

Throughout this book, readers will explore different types of datasets, algorithms, methodologies, and interestingness metrics used in pattern mining. While the book focuses primarily on mining certain data, it also touches upon emerging trends and innovations, such as pattern mining in uncertain data and integration with machine learning techniques.

Whether you are a student, researcher, data scientist, or industry practitioner, this book aims to be a valuable resource. It provides theoretical insights and practical guidance on effectively navigating the complexities of pattern mining. This book also guides researchers in evaluating the algorithms, plotting the results, and generating the latex files for publication purposes. We hope this book serves as a beacon of knowledge, empowering readers to unlock the hidden treasures buried in their data.

Happy mining!

Aizu-Wakamatsu, Fukushima, Japan
March 2025

Uday Kiran Rage

Acknowledgments

This book would not have been possible without the support and guidance of many remarkable individuals and institutions.

First, I would like to express my deepest gratitude to my supervisor, Prof. P. Krishna Reddy, whose wisdom, insight, and constant encouragement were invaluable throughout this journey. His thoughtful mentorship and dedication to my growth pushed me to achieve more than I ever thought possible.

To my family, your unwavering support and love have been the foundation upon which I built this work. Thank you for your patience, understanding, and belief in me throughout this journey. I am deeply grateful for everything you have done.

I would also like to acknowledge the University of Aizu for providing an enriching learning environment, growth, and inspiration. The opportunities, resources, and sense of community I experienced here have shaped this book and played a crucial role in my academic and personal development.

A special thank you to my students, Palla Likhitha, Tarun Sreepada, Suzuki Shota, and Kattumuri Vanitha, whose coding, testing, and proofreading assistance was invaluable. Your hard work, attention to detail, and commitment greatly enhanced the quality of this book, and I sincerely appreciate all your efforts.

Thanks to everyone who contributed to this project in any way. Your support, encouragement, and expertise have made this work possible.

Competing Interests The author has no competing interests to declare that are relevant to the content of this manuscript.

Contents

Part I
Fundamentals

Chapter 1
Getting Started with PAMI: Introduction, Maintenance, and Usage

Abstract Pattern mining is essential for uncovering valuable patterns hidden in big data. While software such as WEKA, Mahout, SPMF, and Knime offer some capabilities, they are often limited in algorithms or integration. To overcome these limitations, researchers at the University of Aizu have developed the pattern mining (PAMI) package. This open-source Python package, available on GitHub and distributed through the Python Package Index, offers over 80 algorithms to identify user interest-based patterns in various databases across multiple computing environments. This chapter introduces the architecture and systematic organization of the algorithms in PAMI. It provides detailed guidance on the installation, maintenance, and execution of the algorithms in PAMI, both from the terminal and within Python programs. Additionally, the chapter explains the input and output requirements for the algorithms, including how they report runtime and memory usage. Through practical examples and instructions, this chapter aims to help users effectively utilize the PAMI package for pattern mining tasks.

1.1 Origins

Pattern mining is a crucial big data analytical technique to uncover interesting patterns hidden in the data. This technique has numerous real-world applications. For example, in market basket analysis, pattern mining helps businesses understand which products are frequently purchased together, supporting inventory management and targeted marketing. In cybersecurity, pattern mining is crucial for anomaly detection, enabling the identification of unusual patterns that could indicate fraudulent activities or security breaches. In transportation systems, pattern mining assists in identifying frequently congested road segments, which is valuable for urban planning and optimizing route recommendations.

Existing pattern mining tools like WEKA [1], Mahout [2], Knime [3], Rapid-Miner [4], MLxtend [5], and Orange [6] typically offer limited algorithms. While these tools are helpful, they may not adequately address the diverse needs of users working with various data types and analytical tasks. On the other hand, specialized software like Coron [7] and LUCS-KDD [8] provides a broader array of algorithms

U. K. Rage, *Hands-on Pattern Mining*,
https://doi.org/10.1007/978-981-96-6791-8_1

tailored for more specific pattern mining tasks. These tools offer greater flexibility and capability in handling complex data mining needs. However, they often come with challenges like outdated development, limited customization, and restricted commercial use.

The Sequence Pattern Mining Framework (SPMF) [9] stands out for its comprehensive algorithms for discovering useful patterns from various databases, including transactional, sequential, and graph databases. However, its Java-based implementation can raise challenges in integrating with popular Python-based machine learning libraries like TensorFlow, PyTorch, and Scikit-learn, which are commonly used in data science.

To address the limitations of current pattern mining tools, researchers at the University of Aizu have developed the **PAttern MIning (PAMI)** [10] package. This open-source Python package, licensed under the GNU V3 License, provides over 80 algorithms for identifying user interest-based patterns across various databases and computing environments.

! Attention

The open-source PAMI package is supplied under the GNU V3 License.

PAMI is designed to be cross-platform, working seamlessly on Windows, Linux, and macOS. It is hosted on GitHub,[1] which fosters transparency and encourages collaborative development. Users can install, update, or uninstall the package via the Python Package Index[2] using the `pip` command.

For detailed guidance, PAMI offers comprehensive documentation on *Read the Docs*,[3] which covers its features, installation, and usage. Additionally, PAMI includes practical examples in Jupyter Notebooks, which can be run on platforms like Google Colab or local machines. These interactive notebooks help users of all skill levels learn and experiment with PAMI's extensive pattern mining capabilities.

1.2 Architecture

The PAMI package adheres to camel casing naming conventions and organizes its algorithms using a hierarchical structure. This systematic arrangement aids in navigating and retrieving algorithms based on their characteristics and functions. The hierarchical structure is outlined as follows:

[1] The PAMI package's source code can be found at https://github.com/UdayLab/PAMI.
[2] The distribution URL of the PAMI package is https://pypi.org/project/pami/.
[3] The URL for code documentation is https://pami-1.readthedocs.io/en/latest/.

1. **Package Name:** The top-level category is "PAMI," encompassing all algorithms within the package.
2. **Theoretical Model:** Algorithms are categorized based on their theoretical models, such as frequent, correlated, and high utility patterns.
3. **Pattern Type:** This level specifies the type of patterns discovered by the algorithms, including:
   - **Basic:** Patterns fitting the given theoretical model
   - **Maximal Patterns:** Patterns that are not subsets of any other patterns
   - **Closed Patterns:** Patterns where no superset has the same support count
   - **Top-*k* Patterns:** Patterns based on frequency, correlation, or periodicity criteria
4. **Mining Algorithms:** The lowest level lists the specific mining algorithms used to extract patterns based on the previous categories.

Figure 1.1a shows an abstract representation of the hierarchical arrangement, while Fig. 1.1b provides a concrete example of the organization within this framework. Additionally, PAMI includes an "extras" sub-package with additional tools

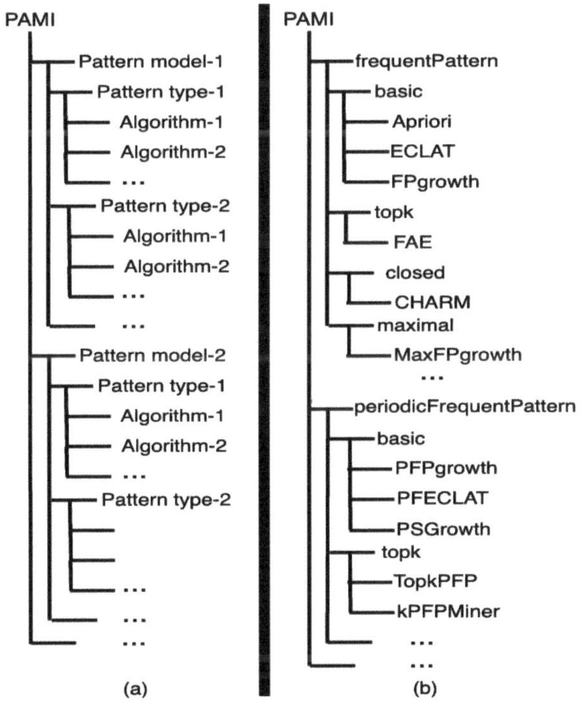

Fig. 1.1 Package structure of PAMI. (**a**) Abstract representation. (**b**) An example

for generating synthetic databases, converting dataframes into various database formats, getting the statistical information of the databases, visualizing the results, and exporting the results in latex for publication purposes.

1.3 Inputs and Outputs of a Mining Algorithm

Figure 1.2 outlines the inputs and outputs of a mining algorithm in the PAMI package. Each algorithm requires data in a specific format and constraints. Data can be provided in three formats: a text file, a dataframe, or a URL for remote datasets. Outputs include the discovered patterns, which can be exported as a list, dataframe, or text file. Algorithms also report their runtime and memory consumption, measured as resident set size (RSS) and unique set size (USS).

1.4 Maintaining the PAMI Package

The PAMI package is designed for easy installation and maintenance using the "pip" command. Table 1.1 lists the basic commands required to manage the package, including installation, upgrading, uninstallation, and showing information.

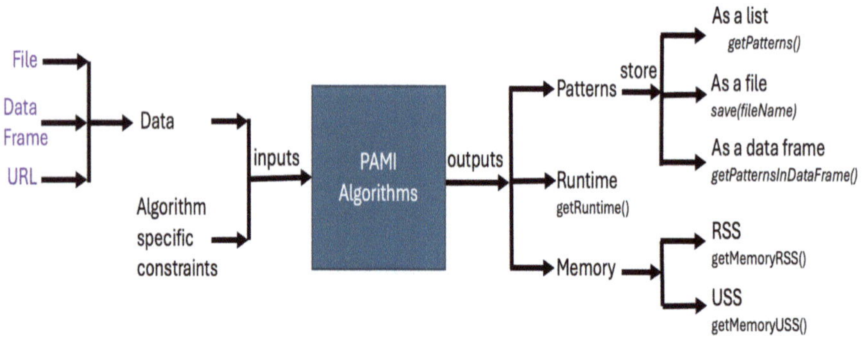

Fig. 1.2 Inputs and outputs of a mining algorithm

Table 1.1 Basic pip commands to maintain PAMI package

S. No.	Purpose	Command
1	Installation	pip install pami
2	Upgradation	pip install –upgrade pami
3	Uninstall	pip uninstall pami
4	Information	pip show pami

1.5 Execution of Algorithms

The algorithms in the PAMI package are executed on a terminal or integrated into their Python programs. We now explain these two processes briefly.

1.5.1 Terminal Execution

To run an algorithm from the terminal, navigate to the algorithm's directory and execute the script with the necessary input and output files and any additional parameters. Below is the generic syntax and a detailed example of executing the renowned Apriori algorithm.

Generic Code 1: Terminal Execution

```
$ cd <pathToAlgorithm>
$ python algorithm.py inputFile outputFile constraints
```

Here:

- <pathToAlgorithm> is the directory containing the algorithm script.
- `algorithm.py` is the script to execute.
- `inputFile` is the file with input data.
- `outputFile` is where the results will be saved.
- `constraints` are additional parameters the algorithm requires.

Example 1: Apriori Execution

```
$ cd PAMI/frequentPattern/basic
$ python Apriori.py sampleDB.txt patterns.txt 10
```

In the above example:

1. **Change Directory**: `$ cd PAMI/frequentPattern/basic` sets the current directory.
2. **Execute Python Script**: `$ python Apriori.py sampleDB.txt patterns.txt 10` runs the script with:
   - `python`: Invokes the Python interpreter.
   - `Apriori.py`: The script to execute.

- `sampleDB.txt`: Input data file.
- `patterns.txt`: Output file for storing the patterns.
- `10`: Minimum support in count.

This method of running scripts is commonly used in various data processing tasks. Scripts typically take input files, process the data based on defined constraints or parameters, and produce output files containing the results. This approach ensures the workflow is organized and the results are systematically stored for further analysis or reporting.

1.5.2 Importing an Algorithm

Users can import the necessary PAMI modules into their Python programs for more advanced usage. This method not only enables the execution of algorithms but also allows for greater control over data preprocessing, post-processing, visualization, and the incorporation of additional logic or functionality. The generic Python code and an example[4] to implement any mining algorithm from the PAMI package are shown below:

Generic Code 2: Implementing a Pattern Mining Algorithm

```
from PAMI.theoreticalModel.patternType import algorithm as alg
# Initialization
obj = alg.algorithm(inputFile, constraints, sep='\t')

# Mining the patterns
obj.mine()

# Save the discovered patterns
obj.save(outputFileName)

# Print the results
print('Total number of patterns: ' +
    str(len(obj.getPatterns())))
print('Runtime: ' + str(obj.getRuntime()))
print('Memory (RSS): ' + str(obj.getMemoryRSS()))
print('Memory (USS): ' + str(obj.getMemoryUSS()))
```

[4] The file used in this experiment can be downloaded from the URL: https://web-ext.u-aizu.ac.jp/~udayrage/datasets/transactionalDatabases/Transactional_T10I4D100K.csv.

Example 2: Implementing the Apriori Algorithm

```
import PAMI.frequentPattern.basic.Apriori as alg

# Create an Apriori object
obj = alg.Apriori(iFile = 'Transactional_T10I4D100K.csv',
    minSup = 500)
# Run the mining process
obj.mine()
# Save the frequent patterns to an output file
obj.save(oFile = 'patterns.txt')
# Print the results
print('Total number of patterns: ' +
↪ str(len(obj.getPatterns())))
print('Runtime: ' + str(obj.getRuntime()))
print('Memory (RSS): ' + str(obj.getMemoryRSS()))
print('Memory (USS): ' + str(obj.getMemoryUSS()))
```

1.6 Evaluating Multiple Pattern Mining Algorithm

The "PAMI" package helps us evaluate the performance of multiple pattern mining algorithms on a dataset. The generic and the sample codes are provided below:

Generic Code 3: Evaluating Multiple Algorithms

```
#import the algorithms
from PAMI.theoreticalModel.patternType import algorithm1 as
↪ alg1
from PAMI.theoreticalModel.patternType import algorithm2 as
↪ alg2
#you can import multiple algorithms
import pandas as pd #to store results

#Create a list of threshold values
constraintList = [100, 150,200]

#Create a dataframe to store results
```

```
resultDF = pd.DataFrame(columns=['algorithm', 'minSup','patterns', 'runtime', 'memoryRSS', 'memoryUSS'])

#implement each algorithm and store the results in a dataframe

for constraint in constraintList:
    obj = alg1.algorithm1(inputParameters)
    obj.mine()
    resultDF.loc[resultDF.shape[0]]=['algorithm1', constraint,len(obj.getPatterns()), obj.getRuntime(), obj.getMemoryRSS(), obj.getMemoryUSS()]

for constraint in constraintList:
    obj = alg2.algorithm2(inputParameters)
    obj.mine()
    resultDF.loc[resultDF.shape[0]]=['algorithm2', constraint,len(obj.get patterns()), obj.getRuntime(), obj.getMemoryRSS(), obj.getMemoryUSS()]

#repeat the above steps for the remaining algorithms
```

Example 3: Evaluating Multiple Algorithms

```
from PAMI.frequentPattern.basic import Apriori as alg1
from PAMI.frequentPattern.basic import FPGrowth  as alg2
import pandas as pd

minimumSupportCountList = [1000, 1500, 2000, 2500, 3000]

resultDF = pd.DataFrame(columns=['algorithm', 'minSup','patterns', 'runtime', 'memoryRSS', 'memoryUSS'])


for minSupCount in minimumSupportCountList:
    obj = alg1.Apriori(iFile='Transactional_T10I4D100K.csv', minSup=minSupCount,sep='\t')
    obj.mine()
    resultDF.loc[resultDF.shape[0]]=['Apriori', minSupCount,len(obj.getPatterns()), obj.getRuntime(), obj.getMemoryRSS(), obj.getMemoryUSS()]

```

```
for minSupCount in minimumSupportCountList:
    obj = alg2.FPGrowth(iFile='Transactional_T10I4D100K.csv',
      minSup=minSupCount, sep='\t')
    obj.mine()
    resultDF.loc[resultDF.shape[0]]=['FPgrowth',
      minSupCount,len(obj.getPatterns()),
      obj.getRuntime(),obj.getMemoryRSS(),
      obj.getMemoryUSS()]

resultDF #print dataframe
```

1.7 Plotting the Results

The generated results about the number of produced patterns, runtime, and memory can be visualized and exported as graphs using the PAMI library. The generic syntax and a sample code can be found below.

Generic Code 4: Viewing the Results

```
from PAMI.extras.graph import PlotLineGraphs4DataFrame as dif
# Pass the result data frame to the class
obj = dif.PlotLineGraphs4DataFrame(resultDF)
# Plotting the graphs
obj.plot(result=resultDF, xaxis='constraint', yaxis='patterns',
  label='algorithm')
obj.plot(result=resultDF, xaxis='constraint', yaxis='runtime',
  label='algorithm')
obj.plot(result=resultDF, xaxis='constraint',
  yaxis='memoryRSS', label='algorithm')
obj.plot(result=resultDF, xaxis='constraint',
  yaxis='memoryUSS', label='algorithm')
#saving the graphs' results
obj.save(result=resultDF, xaxis='constraint', yaxis='patterns',
  label='algorithm',oFile='patterns.jpg')
obj.save(result=resultDF, xaxis='constraint', yaxis='runtime',
  label='algorithm',oFile='runtime.jpg')
```

```python
obj.save(result=resultDF, xaxis='constraint', yaxis='memoryRSS', label='algorithm',oFile='memoryRSS.jpg')
obj.save(result=resultDF, xaxis='constraint', yaxis='memoryUSS', label='algorithm',oFile='memoryUSS.jpg')
```

Example 4: Viewing the Results

```python
from PAMI.extras.graph import PlotLineGraphs4DataFrame as dif
# Pass the result data frame to the class
obj = dif.PlotLineGraphs4DataFrame(resultDF)
# Draw the graphs
obj.plot(result=resultDF, xaxis='minSup', yaxis='patterns', label='algorithm')
obj.plot(result=resultDF, xaxis='minSup', yaxis='runtime', label='algorithm')
obj.plot(result=resultDF, xaxis='minSup', yaxis='memoryRSS', label='algorithm')
obj.plot(result=resultDF, xaxis='minSup', yaxis='memoryUSS', label='algorithm')
#saving the graphs' results
obj.save(result=resultDF, xaxis='minSup', yaxis='patterns', label='algorithm',oFile='patterns.jpg')
obj.save(result=resultDF, xaxis='minSup', yaxis='runtime', label='algorithm',oFile='runtime.jpg')
obj.save(result=resultDF, xaxis='minSup', yaxis='memoryRSS', label='algorithm',oFile='memoryRSS.jpg')
obj.save(result=resultDF, xaxis='minSup', yaxis='memoryUSS', label='algorithm',oFile='memoryUSS.jpg')
```

1.8 Exporting the Results in Latex Format

The "extras" package in the PAMI library contains a Python program that accepts the dataframe containing the results of various algorithms and outputs the latex code that the researchers can later use in their experimental section to draw plots. The generic Python code and an example to export the results in the Latex format are shown below:

Generic Code 5: Exporting the Results in Latex

```
from PAMI.extras.graph import Results2Latex as res
#Initailize
obj = res.Results2Latex()

#Printing the latex code
obj.print(result=resultDF,xaxis='xLabel',yaxis='yLabel',\
    label='algorithm')
#Saving the latex code in a file
obj.save(result=resultDF,xaxis='xLabel',yaxis='yLabel',\
    label='algorithm',oFile='outputFileName.txt')
```

Example 5: Exporting the Results in Latex

```
from PAMI.extras.graph import Results2Latex as res

obj = res.Results2Latex()
#Printing the latex code on the terminal
obj.print(result=resultDF, xaxis='minSup',
↪ yaxis='patterns',label='algorithm')
obj.print(result=resultDF, xaxis='minSup', yaxis='runtime',
↪ label='algorithm')
obj.print(result=resultDF, xaxis='minSup',
↪ yaxis='memoryRSS',label='algorithm')
obj.print(result=resultDF, xaxis='minSup', yaxis='memoryUSS',
↪ label='algorithm')
#save the latex code in a file
obj.save(result=resultDF, xaxis='minSup', yaxis='patterns',
↪ label='algorithm', oFile='patterns.txt')
obj.save(result=resultDF, xaxis='minSup', yaxis='runtime',
↪ label='algorithm', oFile='runtime.txt')
obj.save(result=resultDF, xaxis='minSup', yaxis='memoryRSS',
↪ label='algorithm', oFile='memoryRSS.txt')
obj.save(result=resultDF, xaxis='minSup', yaxis='memoryUSS',
↪ label='algorithm', oFile='memoryUSS.txt')
```

1.9 Contributing

We welcome contributions to the PAMI package. If you have suggestions, improvements, or bug fixes, please mention them in the Discussions section of GitHub. Your contributions are crucial for enhancing the package and supporting the community.

1.10 Support

For support and troubleshooting, please check the Issues section of the GitHub repository. If you encounter any specific problems or need further assistance, contact the maintainers directly through the Discussion Forum.

1.11 Conclusion

The PAMI package has been designed and developed to empower users with robust tools for discovering and analyzing patterns in their data. Whether you are conducting market research, analyzing transactional data, or exploring new trends, this package provides the functionalities needed to perform these tasks efficiently. We encourage you to explore its features, utilize its capabilities, and integrate it into your data analysis workflows.

References

1. Frank, E., Hall, M. A., Holmes, G., Kirkby, R., Pfahringer, B., Witten, I. H. 2005. Weka: A machine learning workbench for data mining. In: *Data Mining and Knowledge Discovery Handbook: A Complete Guide for Practitioners and Researchers*, Maimon, O. and Rokach, L. (ed), 1305–1314, Berlin: Springer publications.
2. Apache software foundation. Mahout. https://mahout.apache.org//, 2020. [Online accessed 13-March-2025].
3. Michael R. Berthold, Nicolas Cebron, Fabian Dill, Thomas R. Gabriel, Tobias Kötter, Thorsten Meinl, Peter Ohl, Kilian Thiel, and Bernd Wiswedel. 2009. KNIME - the Konstanz information miner: version 2.0 and beyond. SIGKDD Explor. Newsl. 11, 1 (June 2009), 26–31.
4. Ralf Klinkenberg, Ingo Mierswa, and Simon Fischer. RapidMiner. https://rapidminer.com/, 2001. [Online accessed 13-March-2025].
5. Sebastian Raschka. MLxtend: Providing machine learning and data science utilities and extensions to Python's scientific computing stack. In: *the Journal of Open Source Software*, 3(24), April 2018.
6. Ferenc Borondics. Orange data mining. https://orangedatamining.com/, 1996. [Online accessed 13-March-2025].
7. Laszlo SZATHMARY, Amedeo NAPOLI, Yannick TOUSSAINT. Coron data mining. http://coron.loria.fr/site/index.php, 2007. [Online accessed 31-August-2022].

8. Frans Coenen. LUCS-KDD. https://cgi.csc.liv.ac.uk/~frans/KDD/Software/, 2013. [Online accessed 13-March-2025].
9. Philippe Fournier-Viger, Antonio Gomariz, Ted Gueniche, Azadeh Soltani, Cheng-Wei Wu, and Vincent S. Tseng. SPMF: A java open-source pattern mining library. In: *J. Mach. Learn. Res.*, vol. 15(1):3389–3393, jan 2014.
10. Uday Kiran Rage, Veena Pamalla, Masashi Toyoda, Masaru Kitsuregawa. PAMI: An Open-Source Python Library for Pattern Mining. In: *J. Mach. Learn. Res.*, vol. 25(209), 1–6.

Chapter 2
Handling Big Data: Classification, Storage, and Processing Techniques

Abstract This chapter offers a comprehensive overview of big data, focusing on its classification, storage, and processing. The first section explores different categories of big data based on structure, veracity, attribute values, dimensionality, and storage. The second section examines popular data storage mechanisms, such as files and database management systems, and provides Python code for converting data between CSV and Parquet formats. The third section discusses data processing using Pandas DataFrames, highlighting their strengths and limitations. The chapter concludes with a summary, providing essential insights for managing and analyzing big data effectively.

2.1 Basic Classifications of Big Data

Many real-world applications naturally produce big data. This data is characterized by volume, velocity, and variety. The term "Big data" represents a wide range of diverse datasets generated by the combinations of various factors such as structure, veracity, an object's values, dimensionality, and storage. Below, we briefly explain the fundamental forms of big data.

2.1.1 Based on Data Structure

1. **Structured Data**: Data organized in a fixed format, such as databases and spreadsheets. Examples include financial records and customer information.
2. **Unstructured Data**: Data that lacks a predefined structure, such as text files, multimedia files, and social media content.
3. **Semi-structured Data**: Data that does not conform to a fixed structure but contains tags to separate data elements. Examples include JSON and XML files.
4. **Graph Data**: Data represented in the form of graphs, where entities are depicted as nodes (vertices), and the relationships between these entities are depicted as edges (links).

U. K. Rage, *Hands-on Pattern Mining*,
https://doi.org/10.1007/978-981-96-6791-8_2

2.1.2 Based on Veracity

1. **Certain Data**: It is the data that is highly reliable, accurate, and complete. This data is typically collected from trusted sources with rigorous quality control measures, making it ideal for critical decision-making processes.
2. **Uncertain Data**: It represents the data that may be incomplete, inconsistent, or subject to significant variability. Handling uncertain data requires advanced cleaning, validation, and analysis techniques to mitigate the risks associated with its use.

2.1.3 Based on the Attribute's Value

1. **Binary Data**: Data in which attributes can take on one of two possible values, often represented as 0 and 1, but can also be true/false, yes/no, on/off.
2. **Nonbinary Data**: Data in which attributes can take more than two values, including categorical, ordinal, interval, and ratio data. This is also known as multivalued or continuous data.

2.1.4 Based on the Attribute's Dimensionality

1. **Transactional Data**: Data containing unordered transactions or itemsets[1]
2. **Temporal Data**: Data containing transactions ordered by time
3. **Spatial Data**: Data in which objects are associated with spatial information, such as pixels, points, lines, and polygons

2.1.5 Based on Storage

1. **Databases**: Static data that can be scanned multiple times
2. **Data Streams**: Continuous data flows that can be scanned only once in real time

Understanding these categories is crucial for users to identify the type of data they are working with and the patterns that can be discovered. Different data types and structures significantly impact the methods and algorithms for analysis and pattern mining. For instance, Fig. 2.1 shows that combining the "structured," "certain," "binary," "transactional," and "database" factors results in the generation of a "structured certain binary transactional database" (or simply, transactional database), while combining "structured," "certain," "binary," "transactional," and "stream" results in a "structured certain binary transactional stream" (or simply,

[1] In pattern mining, a set of items is often written as itemset instead of item set.

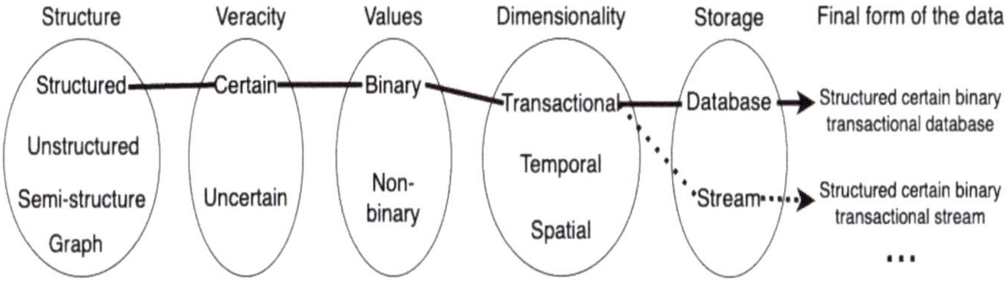

Fig. 2.1 Representing the real-world data

transactional stream). This understanding is crucial as pattern mining models and algorithms designed for handling a particular data type, say transactional databases, may suffer from correctness issues when applied to other data types, such as transactional streams.

2.2 Approaches for Storing Big Data

Big data is widely stored as files due to its simplicity in creation and sharing. Most public data is saved and shared in various file formats, such as Comma Separated Value (CSV), Joint Photographic Experts Group (JPEG), Tag Image File Format (TIFF), Network Common Data Form (NetCDF), Avro, Parquet, and Optimized Row Columnar (ORC). Despite their popularity, files suffer from data integrity, consistency, redundancy, and security issues. To tackle these problems, companies store their big data using database management systems (DBMS) such as *relational databases* (e.g., MySQL [1] and PostgreSQL [2]), *NoSQL databases* (e.g., Apache Cassandra [3] and MongoDB [4]), and *newSQL databases* (e.g., Google Spanner [5] and CockroachDB [6]).

Due to the heterogeneity and the complexity of writing generic code for various DBMS, we confined the algorithms in PAMI to reading the input data as a text file for simplicity. Currently, the PAMI package provides scripts to convert a CSV file into a Parquet and vice versa. The generic code for file conversions is provided below.

Generic Code 1: Converting a CSV File into a Parquet File

```
from PAMI.extras.convert import CSV2Parquet as alg

obj = alg.CSV2Parquet(inputFile,outputFile,sep)
obj.convert()

print('Runtime: ' + str(obj.getRuntime()))
print('Memory (RSS): ' + str(obj.getMemoryRSS()))
print('Memory (USS): ' + str(obj.getMemoryUSS()))
```

Example 1: CSV File to Parquet File

```
import PAMI.extras.convert.CSV2Parquet as cp

obj = cp.CSV2Parquet(inputFile='Transactional_T10I4D100K.csv',\
      outputFile='Transactional_T10I4D100K.parquet',sep='\t')
obj.convert()

print('Runtime: ' + str(obj.getRuntime()))
print('Memory (RSS): ' + str(obj.getMemoryRSS()))
print('Memory (USS): ' + str(obj.getMemoryUSS()))
```

Generic Code 2: Converting a Parquet File into a CSV File

```
from PAMI.extras.convert import Parquet2CSV as alg

obj = alg.Parquet2CSV(inputFile,outputFile,sep)
obj.convert()

print('Runtime: ' + str(obj.getRuntime()))
print('Memory (RSS): ' + str(obj.getMemoryRSS()))
print('Memory (USS): ' + str(obj.getMemoryUSS()))
```

Example 2: Parquet File to CSV

```
import PAMI.extras.convert.Parquet2CSV as cp

obj = cp.Parquet2CSV(inputFile='Transactional_T10I4D100K.
      parquet',\
      outputFile='new_Tran_T10I4D100K.csv',sep='\t')
obj.convert()

print('Runtime: ' + str(obj.getRuntime()))
print('Memory (RSS): ' + str(obj.getMemoryRSS()))
print('Memory (USS): ' + str(obj.getMemoryUSS()))
```

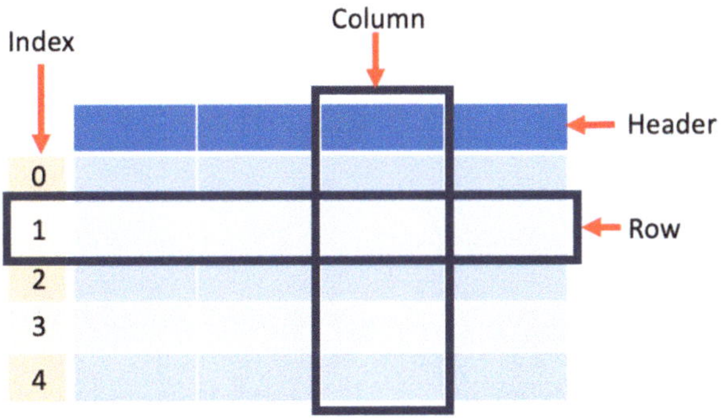

Fig. 2.2 Elements of a dataframe

2.3 Processing Big Data

A DataFrame is a data structure that organizes data into a two-dimensional table (see Fig. 2.2) of rows and columns, much like a spreadsheet. It is one of the most common data structures used in modern data analytics because of its flexibility and intuitive way of storing and working with data. Popular libraries include Pandas DataFrame, Apache Spark DataFrame, Dask DataFrame, and Koalas. Pandas DataFrame is widely used due to its flexibility and powerful data structure for big data processing, particularly for data manipulation and analysis. However, Pandas has a limitation in handling large datasets that exceed memory capacity. Users can address this problem by employing Apache Spark DataFrame, which can handle larger-than-memory datasets by parallelizing operations and distributing data across multiple machines.

Currently, the algorithms in PAMI support Pandas DataFrame and Spark DataFrame. In particular, the sequential algorithms in PAMI support Pandas DataFrame, while the distributed algorithms based on the map-reduce framework support Spark DataFrame.

The PAMI package provides several scripts to convert a DataFrame into a specific database format. Below are the generic Python code and an example.

Generic Code 3: Converting a Dataframe into a Particular Database

```
from PAMI.extras.convert import DF2DB as alg
import pandas as pd
import numpy as np
obj = alg.DF2DB(dataFrame)
obj.convert2ParticularDatabase(outputFileName, other
    parameters)
print('Runtime: ' + str(obj.getRuntime()))
```

```
print('Memory (RSS): ' + str(obj.getMemoryRSS()))
print('Memory (USS): ' + str(obj.getMemoryUSS()))
```

Example 3: Converting a Dataframe into a Transactional Database

```
from PAMI.extras.convert import DF2DB as alg
import pandas as pd
import numpy as np
data = np.random.randint(1, 100, size=(1000, 4))
dataFrame = pd.DataFrame(data, columns=['Item1', 'Item2',
↪   'Item3', 'Item4'])
obj = alg.DF2DB(dataFrame)
obj.convert2TransactionalDatabase(oFile='transactionalDB.csv',
↪   condition='>=', thresholdValue=36)
print('Runtime: ' + str(obj.getRuntime()))
print('Memory (RSS): ' + str(obj.getMemoryRSS()))
print('Memory (USS): ' + str(obj.getMemoryUSS()))
```

2.4 Conclusion

This chapter offers a detailed exploration of big data, from its classification and storage methods to its processing techniques. Understanding these aspects empowers the readers to select the right pattern mining model and an appropriate mining algorithm for knowledge discovery.

References

1. Michael Monty Widenius, David Axmark, and Allan Larsson. MySQL. https://www.mysql.com/, 1995. [Online accessed 13-March-2025].
2. Michael Stonebraker. PostgreSQL. https://www.postgresql.org/, 1986. [Online accessed 13-March-2025].
3. Avinash Lakshman and Prashant Malik. Apache Cassandra. https://cassandra.apache.org/_/index.html, 2008. [Online accessed 13-March-2025].
4. Dwight Merriman, Eliot Horowitz, and Kevin Ryan. MongoDB. https://www.mongodb.com/, 2007. [Online accessed 13-March-2025].
5. James C. Corbett, Jeffrey Dean, Michael Epstein, Andrew Fikes, Christopher Frost, JJ Furman, Sanjay Ghemawat, Andrey Gubarev, Christopher Heiser, Peter Hochschild, Wilson

Hsieh, Sebastian Kanthak, Eugene Kogan, Hongyi Li, Alexander Lloyd, Sergey Melnik, David Mwaura, David Nagle, Sean Quinlan, Rajesh Rao, Lindsay Rolig, Dale Woodford, Yasushi Saito, Christopher Taylor, Michal Szymaniak, Ruth Wang. Spanner: Google's Globally-Distributed Database. In: *OSDI*, 2012.

6. Rebecca Taft, Irfan Sharif, Andrei Matei, Nathan VanBenschoten, Jordan Lewis, Tobias Grieger, Kai Niemi, Andy Woods, Anne Birzin, Raphael Poss, Paul Bardea, Amruta Ranade, Ben Darnell, Bram Gruneir, Justin Jaffray, Lucy Zhang, and Peter Mattis. 2020. CockroachDB: The Resilient Geo-Distributed SQL Database. In: *Proceedings of the 2020 ACM SIGMOD International Conference on Management of Data (SIGMOD '20)*. Association for Computing Machinery, New York, NY, USA, 1493–1509.

Chapter 3
Transactional Databases: Representation, Creation, and Statistics

Abstract This chapter delves into the concept and practicalities of transactional databases, which are crucial for managing and analyzing data across various fields. A transactional database consists of an unordered collection of transactions, each comprising a set of items represented in binary form. The chapter begins with a formal definition of transactional databases using set theory, explaining transactions and patterns. It then addresses practical aspects, including how these databases are stored and formatted on computing devices, with specific file creation and management guidelines. The chapter also covers methods for generating synthetic transactional databases for testing and benchmarking purposes, converting structured dataframes into transactional databases, and analyzing database statistics, including transaction length, item frequency, and sparsity. Overall, it provides a comprehensive overview of both theoretical and practical aspects of transactional databases, offering valuable data management and analysis insights.

3.1 Introduction

A structured certain binary transactional database, or simply a transactional database, is a collection of unordered transactions. Each transaction consists of items, often represented in binary form to indicate their presence or absence. This data format is prevalent in various real-world scenarios, such as sales, healthcare, clickstream, and sensor networks. Figure 3.1 visualizes how different factors combine to form a transactional database, highlighting the complex relationships involved.

This chapter covers the following key aspects of transactional databases:

1. **Theoretical Representation**: The formal definition of a transactional database using set theory
2. **Practical Representation**: How computer systems implement and store transactional databases
3. **Synthetic Database Creation**: Techniques for generating synthetic transactional databases for testing and benchmarking

U. K. Rage, *Hands-on Pattern Mining*,
https://doi.org/10.1007/978-981-96-6791-8_3

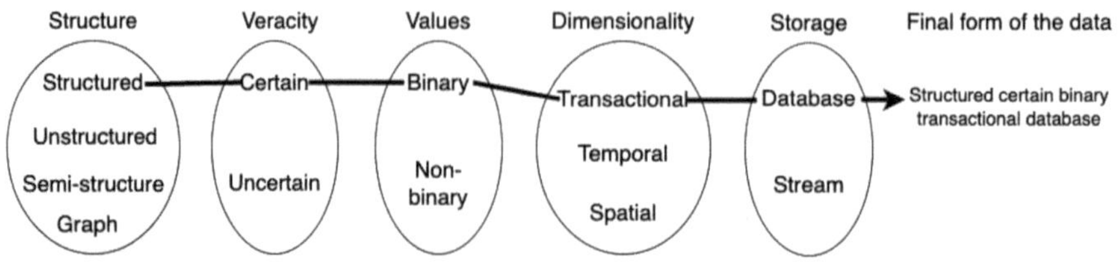

Fig. 3.1 Illustration of factors contributing to the creation of a transactional database

4. **Dataframe Conversion**: Methods to convert structured dataframes into transactional databases for broader data analysis applications
5. **Database Statistics**: How to derive statistical details about a transactional database

3.2 Theoretical Representation

A transactional database [1] is a collection of transactions, each uniquely identified and containing a specific set of items. Formally:

Let $I = \{\text{item}_1, \text{item}_2, \ldots, \text{item}_n\}$, where $n \geq 1$, represent the set of all possible items. An itemset, or **pattern**, is defined as $Y = \{\text{item}_1, \text{item}_2, \ldots, \text{item}_k\} \subseteq I$, where $1 \leq k \leq n$. This subset Y represents a specific combination of items that can occur together in a transaction. A transaction is denoted as $\text{tran} = \{\text{tid}, Y\}$, where $\text{tid} \in \mathbb{R}^+$ is the *transaction identifier*, a unique number for each transaction. The tid ensures distinct transaction identification. The set $Y \subseteq I$ includes the items present in this transaction. A transactional database, denoted as TDB, is a collection of such transactions, formally defined as $TDB = \{\text{tran}_1, \text{tran}_2, \ldots, \text{tran}_m\}$, where $m \geq 1$ represents the total number of transactions in the database.

Example 3.1 Consider the set of items $I = \{\text{Bread, Jam, Butter, Book, Pen}\}$ available in a supermarket. Table 3.1a and b presents the horizontal and vertical formats of a transactional database, respectively. This database is based on the purchases made by five anonymous customers. For simplicity, the concepts will be explained using the horizontal format shown in Table 3.1a.

In the first transaction, $\text{tran}_1 = \{1 : \text{Bread, Jam, Butter}\}$, 1 represents the transaction identifier (or tid), and {Bread, Jam, Butter} represent the items purchased in that transaction. This transaction indicates that a customer purchased the items "Bread," "Jam," and "Butter," uniquely identified by transaction identifier 1. Similar statements can be made about the remaining transactions.

Table 3.1 Hypothetical transactional database of a supermarket

(a) Horizontal format

tid	Items
1	Bread, Jam, Butter
2	Bread, Book, Pen
3	Jam, Butter
4	Bread, Jam, Butter, Pen
5	Book, Pen

(b) Vertical format

tid	Bread	Jam	Butter	Book	Pen
1	1	1	1	0	0
2	1	0	0	1	1
3	0	1	1	0	0
4	1	1	1	0	1
5	0	0	0	1	1

3.3 Practical Representation

A transactional database is usually stored as a file on a computer. To properly create and manage this file, follow these three rules:

- **One Transaction per Line**: Each line in the file represents a single transaction. The line number implicitly acts as the transaction identifier (tid), so it is not explicitly stored in the file to save space and reduce processing costs.
- **Unique Items per Transaction**: Each item should appear only once per line. The items can be listed in any order within the line.
- **Items Separated by a Delimiter**: Items in a transaction are separated by a delimiter, such as a space or tab. The PAMI algorithms use a `tab` as the default delimiter, but users can choose other delimiters like commas or spaces.

Overall, the format of a transaction in a transactional database is:

$$item_1\langle sep\rangle item_2\langle sep\rangle item_3\langle sep\rangle \cdots$$

Example 3.2 If the delimiter is a `tab`, the transactional database shown in Table 3.1a would look like this:

```
Bread	Jam	Butter
Bread	Book	Pen
Jam	Butter
Bread	Jam	Butter	Pen
Book	Pen
```

Important

"Tab" is the default separator the PAMI package uses to distinguish the items within the line of a file.

3.4 Creating Synthetic Transactional Databases

The PAMI package offers a powerful and flexible tool for generating synthetic transactional databases, tailored to various needs. This capability is invaluable for testing and developing algorithms in data mining and related fields. Users can customize the database to suit their specific requirements, including the number of transactions, the total number of items, and the average transaction length.

To illustrate the creation of a synthetic transactional database, consider the following sample code. This example generates a database with 100,000 transactions, each containing an average of 10 items from a set of 1,000 possible items:

Program 1: Generating Synthetic Transactional Database

```
from PAMI.extras.syntheticDataGenerator import TransactionalDatabase as db

obj = db.TransactionalDatabase(
        databaseSize=100000,
        avgItemsPerTransaction=10,
        numItems=1000,
        sep='\t'
        )
obj.create()
obj.save('transactionalDatabase.csv')
#read the generated transactions into a dataframe
transactionalDataFrame=obj.getTransactions()
#stats
print('Runtime: ' + str(obj.getRuntime()))
print('Memory (RSS): ' + str(obj.getMemoryRSS()))
print('Memory (USS): ' + str(obj.getMemoryUSS()))
```

3.5 Deriving a Transactional Database from a Dataframe

The PAMI package enables users to convert a dataframe into a transactional database, which is ideal for transaction-based data analysis. Below is a Python code snippet illustrating how to use PAMI for this conversion:

Program 2: Converting a Dataframe into a Transactional Database

```python
from PAMI.extras.convert import DF2DB as alg
import pandas as pd
import numpy as np

#creating a 1000 x 4 dataframe with random values
data = np.random.randint(1, 100, size=(1000, 4))
dataFrame = pd.DataFrame(data,
             columns=['Item1', 'Item2', 'Item3', 'Item4']
            )
#converting the database into a transactional database by
#considering values greater than or equal to 36
obj = alg.DF2DB(dataFrame)
obj.convert2TransactionalDatabase(oFile='transactionalDB.csv',
       condition='>=', thresholdValue=36
    )
print('Runtime: ' + str(obj.getRuntime()))
print('Memory (RSS): ' + str(obj.getMemoryRSS()))
print('Memory (USS): ' + str(obj.getMemoryUSS()))
```

3.6 Knowing the Statistical Details

The `dbStats` sub-sub-package in the `extras` sub-package of PAMI provides users with statistical details about a transactional database. This functionality is essential for understanding the properties and characteristics of the database, which can be crucial for various data analysis tasks. The statistical details provided by dbStats include:

1. Database size
2. Total number of items in a database
3. Minimum, average, and maximum lengths of the transactions
4. Standard deviation of transactional sizes
5. Variance in transaction sizes

6. Sparsity
7. Frequencies of the items
8. Distribution of transactional lengths

Here is an example of how to use the `dbStats` to obtain the statistics:

Program 3: Deriving the Statistical Details

```
from PAMI.extras.dbStats import TransactionalDatabase as stat

obj = stat.TransactionalDatabase("transactionalDatabase.csv")
obj.run()
obj.printStats()
obj.plotGraphs()
```

3.7 Conclusion

This chapter has provided a comprehensive overview of transactional databases, from their theoretical underpinnings to practical applications. We began with a formal definition of transactional databases, detailing how transactions are structured and identified using set theory. We then explored the practical aspects of how these databases are stored and managed on computing devices, including the rules for formatting and storing transactions.

We discussed methods for generating synthetic transactional databases, which are crucial for testing and benchmarking various pattern mining algorithms. The chapter covered techniques for converting structured dataframes into transactional databases, broadening the data analysis scope. Finally, we examined how to derive and interpret statistical details of transactional databases to understand their properties better and optimize their usage.

Understanding these concepts and techniques equips users with the tools to manage, analyze, and leverage transactional databases in various real-world applications. The combination of theoretical knowledge and practical skills discussed here lays the foundation for advanced data analysis.

Reference

1. Charu C. Aggarwal, Yan Li, Jianyong Wang, and Jing Wang. 2009. Frequent pattern mining with uncertain data. In Proceedings of the 15th ACM SIGKDD international conference on Knowledge discovery and data mining (KDD '09). Association for Computing Machinery, New York, NY, USA, 29–38.

Chapter 4
Pattern Discovery in Transactional Databases

Abstract Useful patterns that can empower the users to achieve socioeconomic development lie hidden in the transactional databases. This chapter introduces various types of user interest-based patterns, such as frequent patterns, correlated patterns, fault-tolerant patterns, and association rules, that can be discovered from transactional databases. This chapter also provides sample Python code to find interesting patterns using the PAMI library.

4.1 Introduction

The previous chapter provided a comprehensive overview of transactional databases, covering their construction, practical representation, and methods for deriving statistical insights. This chapter focuses on the analytical dimension, which involves extracting and analyzing the valuable patterns within the transactional database.

This chapter delves into several critical aspects of mining transactional databases:

1. **Frequent Pattern Discovery**: We will formally define a frequent pattern [1], discuss the search space involved, explain the Apriori property, and outline the other algorithms for discovering these patterns.
2. **Handling Redundancy Problem in Frequent Patterns**: This section addresses the redundancy problem by exploring techniques such as mining closed frequent patterns [3], identifying maximal frequent patterns [2], and selecting top-k frequent patterns [5].
3. **Rare Item Problem**: We will examine the challenges associated with mining frequent patterns containing rare items and discuss their implications.
4. **Solutions to the Problem**: Various strategies to address the rare item problem will be explored, including mining frequent patterns with multiple minimum supports [4], discovering correlated patterns [6, 7], deriving relative frequent patterns [8], and identifying fault-tolerant patterns [9].

U. K. Rage, *Hands-on Pattern Mining*,
https://doi.org/10.1007/978-981-96-6791-8_4

5. **Association Rule Discovery**: Finally, we will cover methods for finding association rules from the discovered frequent patterns.

Chapter 3 introduced the foundational concepts of transactional databases, including key terms such as "pattern," "transaction," and "transactional database." We will continue using these terms consistently throughout this chapter to streamline the discussion and minimize redundancy. For readers who may have missed the previous chapter, we recommend reviewing at least Sect. 3.2 to familiarize themselves with the essential concepts and terminologies.

! Attention

The fundamental concepts of transactional databases were described in Chap. 3.

4.2 Frequent Patterns

Frequent patterns are an important class of regularities that can be identified within transactional databases. They are foundational for discovering additional patterns that reflect user interests and behaviors. This section delves into frequent patterns in detail, emphasizing their basic model, search space, the Apriori property, and procedure for finding them using the PAMI library. Mastery of frequent patterns is essential for uncovering key relationships within the data and is the basis for more advanced pattern mining techniques.

4.2.1 Basic Model

Definition 4.1 (Support of a Pattern) Let $P \subseteq I$ be a pattern. The *support* of P in a transactional database TDB is defined as

$$\text{sup}(P) = \frac{\text{freq}(P)}{|TDB|},$$

where $\text{freq}(P)$ denotes the frequency of pattern P in TDB, and $|TDB|$ represents the total number of transactions in the database.

Example 4.1 Let $\{Bread, Jam, Butter\}$ be a pattern. This pattern appears in two transactions of Table 3.1a. Hence, the $frequency$ of this pattern is 2. The $support$ of this pattern, i.e., $sup(\{Bread, Jam, Butter\}) = \frac{2}{5} = 0.4(= 40\%)$. It means 40% of the customers have purchased the items "Bread," "Jam," and "Butter."

Definition 4.2 (Frequent Pattern) The pattern P is said to be a frequent pattern if $sup(P) \geq minSup$, where $minSup$ represents the user-specified *minimum support*.

Example 4.2 If the user-specified *minimum support* is 30%, i.e., $minSup = 30\%$, then the pattern $\{Bread, Jam, Butter\}$ is considered as a frequent pattern because $sup(\{Bread, Jam, Butter\}) \geq minSup$.

In the pattern mining literature, the terms $frequency$ and $support$ are often used interchangeably for clarity and simplicity. In this book, we have chosen to consistently use the term $support$ to refer to the $frequency$ of a pattern. By doing so, we aim to streamline explanations and ensure consistency throughout the text, making it easier for readers to grasp the underlying concepts without confusion.

Important

This book uses the terms $frequency$ and $support$ interchangeably for brevity.

Definition 4.3 (Problem Definition) Given a transactional database (TDB) and the user-specified *minimum support* ($minSup$) value, discover all frequent patterns in TDB that have $support$ greater than or equal to the user-specified $minSup$ value.

4.2.2 Search Space

The space of items in a transactional database raises an itemset lattice (see Fig. 4.1). This lattice represents the search space of pattern mining. Thus, the search space of frequent pattern mining (or any related pattern mining) is $2^n - 1$, where n represents the total number of items in a database. This vast search space makes pattern mining a nontrivial and challenging task.

Example 4.3 The transactional database shown in Table 3.1 contains five items. The itemset lattice for these five items defines the search space for frequent pattern mining. Consequently, this database's search space size for frequent pattern mining

Fig. 4.1 The itemset lattice of a, b, and c items

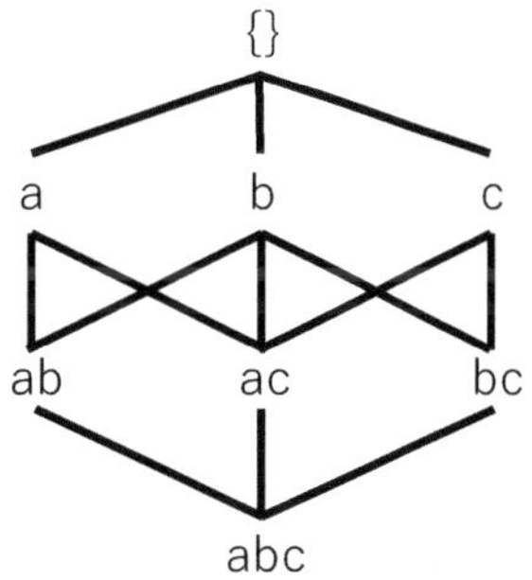

is $2^5 - 1 = 32 - 1 = 31$. In other words, a frequent pattern mining algorithm has to do a traversal among all 31 patterns to find the complete set of frequent patterns.

4.2.3 The Apriori Property

When encountering the problem of enormous search space, the researchers try to tackle it using the *Apriori* (or *downward closure*) property. This property states that "*All non-empty subsets of a frequent pattern must also be frequent.*" This property makes frequent pattern mining practical in real-world applications.

Example 4.4 In Table 3.1, the $sup(\{Bread, Book\}) \geq sup(\{Bread, Book, Pen\})$. If the user-specified $minSup = 40\%$, $\{Bread, Book\}$ is not a frequent pattern as $sup(\{Bread, Book\}) \not\geq minSup$. Furthermore, $\{Bread, Book, Pen\}$ cannot be a frequent pattern as $sup(\{Bread, Book\}) \geq sup(\{Bread, Book, Pen\}) \not\geq minSup$. Thus, we can stop searching all the supersets of $\{Bread, Book\}$ once we discovered that it is not a frequent pattern.

Apriori Property

If a pattern fails a test, its supersets also fail the test.

4.2.4 Finding Frequent Patterns

The literature describes several algorithms for finding frequent patterns, such as Apriori [1], ECLAT [10, 11], and FP-growth [12]. Although no universally acceptable best algorithm exists for finding frequent patterns in any transactional database, most researchers utilize FP-growth as it is generally faster than the other algorithms. Below is a sample Python script for finding frequent patterns using the FP-growth algorithm available in the PAMI package.

Program 1: Frequent Pattern Discovery Using FP-Growth

```
from PAMI.frequentPattern.basic import FPGrowth  as alg

obj = alg.FPGrowth(iFile='Transactional_T10I4D100K.csv',
↪   minSup=300, sep='\t')
obj.mine()
obj.save('frequentPatternsAtMinSupCount300.txt')

frequentPatternsDF= obj.getPatternsAsDataFrame()
```

```
print('#Patterns: ' + str(len(frequentPatternsDF)))
print('Runtime: ' + str(obj.getRuntime()))
print('Memory (RSS): ' + str(obj.getMemoryRSS()))
print('Memory (USS): ' + str(obj.getMemoryUSS()))
```

4.2.5 Popular Variants of Frequent Patterns

Since the objective of the basic frequent pattern model is to find all patterns that satisfy the user-specified *minSup* in a transactional database, it often generates too many patterns, most of which may be redundant or uninteresting depending on the user and application requirements.

Example 4.5 The basic frequent pattern model not only finds $\{Bread, Jam, Butter\}$ as a frequent pattern in Table 3.1a, but also finds all of its non-empty subsets, i.e., $\{Bread\}$, $\{Jam\}$, $\{Butter\}$, $\{Bread, Jam\}$, $\{Bread, Butter\}$, and $\{Jam, Butter\}$, as frequent patterns. Due to redundancy, users may feel these non-empty subsets of $\{Bread, Jam, Butter\}$ uninteresting.

Researchers tried to tackle this problem by finding "maximal frequent patterns," "closed frequent patterns," and "top-*k* frequent patterns." We briefly study these patterns and look at the procedures to find them.

4.2.5.1 Closed Frequent Patterns

A frequent pattern is a closed frequent pattern if none of its supersets have the same support as itself. Suppose FP and CFP, respectively, represent the set of frequent patterns and closed frequent patterns generated from a transactional database at a given $minSup$ value. In that case, their relation is $CFP \subseteq FP$ (or $|CFP| \leq |FP|$). In other words, closed frequent patterns are relatively fewer than the frequent patterns in a database. More importantly, the closed frequent patterns correspond to the lossless representation of frequent patterns, as the complete set of frequent patterns can be regenerated without losing any information from the closed frequent patterns.

Example 4.6 Let us consider the following three frequent patterns in Table 3.1: $\{Bread, Jam\}$, $\{Jam, Butter\}$, and $\{Bread, Jam, Butter\}$. The relation between these three patterns is: $\{Bread, Jam\}$ and $\{Jam, Butter\}$ are the subsets of $\{Bread, Jam, Butter\}$. The $support$ of these patterns, i.e., $sup(\{Bread, Jam\}) = 2$, $sup(\{Jam, Butter\}) = 3$, and $sup(\{Bread, Jam, Butter\}) = 2$. Since the $support$ of $\{Bread, Jam\}$ is the same as that of its superset $\{Bread, Jam, Butter\}$, we can ignore the frequent pattern $\{Bread, Jam\}$ as it can be regenerated from its superset $\{Bread, Jam, Butter\}$ without loss of any information. We cannot say the same for the patterns $\{Jam, Butter\}$ and $\{Bread, Jam, Butter\}$ as both patterns have different $support$ values. Thus, we consider $\{Jam, Butter\}$ and $\{Bread, Jam, Butter\}$ as closed frequent patterns.

The procedure for finding closed frequent patterns in a database is shown below.

Program 2: Finding Closed Frequent Patterns

```
from PAMI.frequentPattern.closed import CHARM  as alg

obj = alg.CHARM(iFile='Transactional_T10I4D100K.csv',
↪  minSup=300)
obj.mine()
obj.save('closedFrequentPatterns.txt')

print('#Patterns: ' + str(len(obj.getPatternsAsDataFrame())))
print('Runtime: ' + str(obj.getRuntime()))
print('Memory (RSS): ' + str(obj.getMemoryRSS()))
print('Memory (USS): ' + str(obj.getMemoryUSS()))
```

Important

Closed frequent patterns denote the lossless representation of frequent patterns.

4.2.5.2 Maximal Frequent Patterns

A frequent pattern is a maximal frequent pattern if none of its supersets are frequent. If FP, CFP, and MFP, respectively, represent the set of frequent patterns, closed frequent patterns, and maximal frequent patterns generated from a transactional database at a given $minSup$ value, then the relation between them is $MFP \subseteq CFP \subseteq FP$ (or $|MFP| \leq |CFP| \leq |FP|$). Unlike closed frequent patterns, maximal frequent patterns correspond to the lossy representation as we cannot derive the exact $support$ information of all the frequent patterns.

Example 4.7 Continuing the previous example, among the closed frequent patterns $\{Jam, Butter\}$ and $\{Bread, Jam, Butter\}$, only $\{Bread, Jam, Butter\}$ is considered as a maximal frequent pattern as none of its supersets represent frequent patterns. We can generate all of its subset frequent patterns from the maximal frequent pattern $\{Bread, Jam, Butter\}$. However, we cannot determine their exact $support$. Henceforth, maximal frequent patterns represent the lossy representation of frequent patterns.

> **Important**
>
> Maximal frequent patterns denote the lossy representation of frequent patterns.

The procedure for finding maximal frequent patterns in a transactional database is below.

Program 3: Finding Maximal Frequent Patterns

```
from PAMI.frequentPattern.maximal import MaxFPGrowth  as alg

obj = alg.MaxFPGrowth(iFile='Transactional_T10I4D100K.csv',
    minSup=300)
obj.mine()
obj.save('maximalFrequentPatternsAtMinSupCount100.txt')

maximalFPsDF= obj.getPatternsAsDataFrame()

print('#Patterns: ' + str(len(obj.getPatternsAsDataFrame())))
print('Runtime: ' + str(obj.getRuntime()))
print('Memory (RSS): ' + str(obj.getMemoryRSS()))
print('Memory (USS): ' + str(obj.getMemoryUSS()))
```

4.2.5.3 Top-*k* Frequent Patterns

The main issue with the basic model of frequent pattern mining is determining the right minimum support value for a transactional database. To tackle this problem, researchers introduced top-*k* frequent pattern mining, where the mining algorithm focuses on finding top-*k* frequently occurring patterns without using the $minSup$ value. The procedure for finding these patterns in a transactional database is below.

Program 4: Finding Top-k Frequent Patterns

```
from PAMI.frequentPattern.topk import FAE  as alg

obj = alg.FAE(iFile='transactionalDatabase.csv', k=1000)
obj.mine()
obj.save('topkFrequentPatterns.txt')
```

```

print('#Patterns: ' + str(len(obj.getPatternsAsDataFrame())))
print('Runtime: ' + str(obj.getRuntime()))
print('Memory (RSS): ' + str(obj.getMemoryRSS()))
print('Memory (USS): ' + str(obj.getMemoryUSS()))
```

4.3 The Rare Item Problem in Frequent Pattern Mining

Since the basic frequent pattern model determines the interestingness of a pattern using only a single $minSup$ value, it implicitly assumes that all items in the database have uniform $support$. However, this is seldom not the case as some items appear frequently, while others appear relatively infrequent (or rarely) in the database. If the $support$ values of the items vary widely in the database, then the frequent pattern model suffers from the following two limitations:

1. If we set a high $minSup$ value, we miss the frequent patterns containing rare items as these items fail to satisfy the increased $minSup$ value.
2. We need to set a low $minSup$ value to find the frequent patterns containing frequent and rare items. However, setting a low $minSup$ may cause a combinatorial explosion, producing too many patterns, most of which may be uninteresting to the user depending on the user or application requirements.

This dilemma is known as the `rare item problem`. The example below illustrates this problem.

Example 4.8 In a supermarket, customers frequently purchase cheap and perishable goods, such as bread and butter. These items are bought often and in large quantities. On the other hand, costly and durable goods, such as wine and whiskey, are purchased less frequently. While not bought as often, these items generate significant revenue when sold. Supermarket managers are often more interested in understanding the purchasing patterns of these rarely bought but high-revenue items. However, due to the rare item problem, it is challenging to discover patterns that include these rare items. If a high $minSup$ value is used, patterns involving wine and whiskey are likely to be missed. If a low $minSup$ value is used, the supermarket managers are overwhelmed with too many patterns, most of which are useless.

4.4 Solutions to the Rare Item Problem

We now discuss some of the famous and widely used solutions presented by the researchers in the literature to tackle the rare item problem.

4.4.1 Finding Frequent Patterns Using Multiple Minimum Supports

In this approach, every item in the database is specified a minimum support-like constraint, known as *minimum item support* (MIS). Next, the minimum support of a pattern is defined as the minimum of its items' MIS values. A pattern is considered frequent if its $support$ is no less than its items' lowest MIS value (see Definition 4.4).

Definition 4.4 (Frequent Pattern) A pattern P is a frequent pattern if $sup(P) \geq min(MIS(i_j)|\forall i_j \in P)$, where $MIS(i_j)$ represents the minimum item support of an item $i_j \in P$. A popular approach to specifying the items' MIS values is the percentage-based methodology, which is as follows:

$$MIS(i_j) = max(sup(i_j) \times \beta, LS), \tag{4.1}$$

where $\beta \in (0, 1)$ is a constant that captures the percentage value, and LS represents the least support a pattern can maintain in the database. The LS parameter removes highly infrequent (or noisy) items in the data.

Example 4.9 Let the $support$ values of the items "Bread," "Butter," "Wine," and "Whiskey" in sales data be 1000, 500, 100, and 60, respectively. Let us set $LS = 40$, i.e., any pattern, irrespective of its (frequent or rare) items, must appear at least 40 times in the data. If $\beta = 0.5$, then $MIS(Bread) = max(0.5 \times 1000, 40) = 500$, $MIS(Butter) = 250$, $MIS(Wine) = 50$, and $MIS(Whiskey) = 40$ $(= max(0.5 \times 60, 40))$. The pattern $\{Bread, Butter\}$ containing the frequently purchased items can be considered frequent if its $support$ is no less than 250 $(= min(500, 250))$. Similarly, the pattern $\{Wine, Whiskey\}$ containing the rarely purchased items can be considered frequent if its $support$ is no less than 40 $(= min(50, 40))$. Thus, depending upon its items, each pattern can satisfy a different $minSup$ value in the multiple minimum support frequent pattern model.

We now examine the procedures for specifying the items' MIS values and finding frequent patterns using multiple minimum supports.

Program 5: Specifying MIS Values for the Items

```python
from PAMI.extras.calculateMISValues import usingBeta as ub
cd = ub.usingBeta(iFile='Transactional_T10I4D100K.csv',
↪  beta=0.5, LS=100) #using default tab separator
cd.calculateMIS()
cd.save('MIS.txt')
```

Program 6: Frequent Pattern Discovery

```
from PAMI.multipleMinimumSupportBasedFrequentPattern.basic import CFPGrowthPlus as alg

obj = alg.CFPGrowthPlus(iFile='Transactional_T10I4D100K.csv', MIS='MIS.txt')  #using default tab separator
obj.mine()
obj.save('frequentPatternsMultipleMinimumSupports.txt')
print('Total No of patterns: ' + str(len(obj.getPatternsAsDataFrame())))
print('Runtime: ' + str(obj.getRuntime()))
print('Memory (RSS): ' + str(obj.getMemoryRSS()))
print('Memory (USS): ' + str(obj.getMemoryUSS()))
```

4.4.2 Correlated Patterns

A significant obstacle to the widespread adoption of frequent pattern mining in real-world applications is its failure to capture the genuine correlation relationship among data objects. Researchers have tried discovering correlated patterns using alternative measures of support to confront the obstacle. Although no universally accepted best measure exists to judge the interestingness of a pattern, *all-confidence* is emerging as a measure that can disclose genuine correlation relationships among data objects. We now define the model of correlated patterns using the *all-confidence* measure.

Definition 4.5 (All-Confidence of a Pattern) The all-confidence of a pattern P, denoted as $all\text{-}conf(P)$, can be expressed as the ratio of its support to the maximum support of an item within it. That is,

$$all\text{-}conf(P) = \frac{sup(P)}{max(sup(i_j)|\forall i_j \in P)}. \tag{4.2}$$

Example 4.10 Consider the frequent pattern $\{Bread, Jam, Butter\}$ in Table 3.1a. The all-confidence of this pattern, i.e.,

$$\begin{aligned} all\text{-}conf(\{Bread, Jam, Butter\}) &= \frac{sup(\{Bread, Jam, Butter\})}{max(sup(Bread), sup(Jam), sup(Butter))} \\ &= \frac{2}{max(3, 3, 3)} \\ &= \frac{2}{3} \\ &= 0.666\ (= 66.6\%). \end{aligned}$$

Definition 4.6 (Correlated Pattern) A frequent pattern P is a correlated pattern if its *all-confidence* value is greater than or equal to the user-specified *minimum all-confidence* ($minAllConf$) value. In other words, P is a correlated pattern if $sup(P) \geq minSup$ and $allConf(P) \geq minAllConf$.

Example 4.11 If the user-specified $minAllConf = 0.5$ (= 50%), the frequent pattern $\{Bread, Jam, Butter\}$ is said to be a correlated pattern because its all-confidence value is greater than or equal to the user-specified $minAllConf$ value.

The search space for correlated pattern mining is the same as that for frequent patterns. The Python script to find the correlated patterns in a transactional database is shown below.

Program 7: Finding Correlated Patterns

```
from PAMI.correlatedPattern.basic import CoMine as alg

obj = alg.CoMine(iFile='Transactional_T10I4D100K.csv',
↪   minSup=300, minAllConf=0.5)
obj.mine()
obj.save('correlatedPatterns.txt')

print('#Patterns: ' + str(len(obj.getPatternsAsDataFrame())))
print('Runtime: ' + str(obj.getRuntime()))
print('Memory (RSS): ' + str(obj.getMemoryRSS()))
print('Memory (USS): ' + str(obj.getMemoryUSS()))
```

4.4.3 Relative Frequent Patterns

Relative frequent patterns are a special type of correlated patterns discovered using the *relative support* measure instead of the *all-confidence* measure. We now define the model of relative frequent patterns.

Definition 4.7 (The *Relative Support* of a Pattern) The relative support of a pattern P, denoted as $RS(P)$, can expressed as the ratio of its support to the minimum support of its items. That is,

$$RS(P) = \frac{sup(P)}{min(sup(i_j)|\forall i_j \in P)}. \tag{4.3}$$

Example 4.12 Consider the frequent pattern $\{Bread, Jam, Butter\}$ in Table 3.1a. The *relative support* of this pattern, i.e.,

$$
\begin{aligned}
RS(\{Bread, Jam, Butter\}) &= \frac{sup(\{Bread, Jam, Butter\})}{min(sup(Bread), sup(Jam), sup(Butter))} \\
&= \frac{2}{min(3, 3, 3)} \\
&= \frac{2}{3} \\
&= 0.666 \ (= 66.6\%).
\end{aligned}
$$

Definition 4.8 (Relative Frequent Pattern P) A frequent pattern P is said to be a relative frequent pattern if $RS(P) \geq minRS$, where $minRS \in (0, 1)$ represents the user-specified *minimum relative support* value.

Definition 4.9 If $minRS = 60\%$, then the frequent pattern {Bread, Jam, Butter} is a relative frequent pattern because $RS(\{Bread, Jam, Butter\}) \geq minRS$.

The Python script to find the relative frequent patterns in a database is provided below.

Program 8: Finding Relative Frequent Patterns

```python
from PAMI.relativeFrequentPattern.basic import RSFPGrowth as
↪ alg

obj = alg.RSFPGrowth(iFile='Transactional_T10I4D100K.csv',
↪ minSup=300, minRS=0.6)

obj.mine()
obj.save('relativeFrequentPatterns.txt')

relativeFrequentPatternsDF= obj.getPatternsAsDataFrame()
print('#Patterns: ' + str(len(relativeFrequentPatternsDF)))
print('Runtime: ' + str(obj.getRuntime())) #measure the runtime
print('Memory (RSS): ' + str(obj.getMemoryRSS()))
print('Memory (USS): ' + str(obj.getMemoryUSS()))
```

4.4.4 *Fault-Tolerant Patterns*

In real-world data mining scenarios, especially in databases with noisy or incomplete data, it is crucial to find patterns that are robust to such imperfections. Fault-tolerant frequent patterns refer to patterns that remain valid even when some of the data items are missing, erroneous, or noisy. This concept is essential for ensuring that discovered patterns are reliable and valuable despite potential data quality issues.

Fault-tolerant frequent itemsets extend traditional frequent itemset mining by incorporating tolerance to missing or incorrect items within transactions. This method modifies the support count mechanism to account for the presence of faults. In particular, a pattern is considered frequent if it appears in most transactions, even if some items within the transactions are missing or incorrect.

Definition 4.10 (Fault-Tolerant Frequent Pattern) The length of pattern P, i.e., $|P| > \gamma$, where $\gamma > 0$ represents the user-specified fault tolerance threshold value. A transaction $Tran = (tid, Y)$ is said to be **FT-containing** pattern P iff there exists $P' \subseteq P$ such that $P' \subseteq Y$ and $|P'| \geq (|P| - \gamma)$. The number of transactions in a database FT-containing pattern P is called the FP-support of P, denoted as $\widehat{sup(P)}$. The pattern P is said to be a fault-tolerant frequent pattern if it satisfies the following two conditions:

1. $\widehat{sup(P)} \geq minSup^{FT}$, where $minSup^{FT}$ represents the user-specified *minimum fault-tolerant support.*
2. For each item $i_j \in P$, $sup(i_j) \geq MIS(i_j)$.

Example 4.13 Consider the pattern $\{Bread, Jam, Butter\}$ in Table 3.1. If the user-specified fault tolerance threshold $\gamma = 1$, then $\{Bread, Jam, Butter\}$ can be considered as a candidate to be a fault-tolerant frequent pattern as $|\{Bread, Jam, Butter\}| \geq 1$. Any two (=3-1) items of the pattern $\{Bread, Jam, Butter\}$ appear in transactions whose $tids$ are 1, 3, and 4. Thus, the FT-support of this pattern is 3. If the user-specified $MIS(Bread) = 2$, $MIS(Jam) = 2$, $MIS(Butter) = 2$, and $minSup^{FT} = 3$, then $\{Bread, Jam, Butter\}$ is a fault-tolerant frequent pattern as $sup(Bread) \geq MIS(Bread)$, $sup(Jam) \geq MIS(Jam)$, $sup(Butter) \geq MIS(Butter)$, and $sup(\{Bread, Jam, Butter\}) \geq minSup^{FT}$.

The Python script to find the fault-tolerant frequent patterns in a database is provided below.

Program 9: Finding Fault-Tolerant Frequent Patterns

```python
from PAMI.faultTolerantFrequentPattern.basic import FTFPGrowth as alg

obj = alg.FTFPGrowth(iFile='Transactional_T10I4D100K.csv', minSup=100, itemSup=100, minLength=3, faultTolerance=1, sep="\t")

obj.mine()
```

```
print('#Patterns: ' + str(len(relativeFrequentPatternsDF)))
print('Runtime: ' + str(obj.getRuntime())) #measure the runtime
print('Memory (RSS): ' + str(obj.getMemoryRSS()))
print('Memory (USS): ' + str(obj.getMemoryUSS()))
```

4.5 Discovering Association Rules

Association rule mining is a popular data mining technique for discovering interesting relationships between the (frequent) patterns in the data. An association rule is of form $A \rightarrow B$, where A and B are patterns such that $A \cap B = \emptyset$. An association rule is interesting if its $confidence$ exceeds the threshold value of the user-specified *minimum confidence* ($minConf$). The $confidence$ of an association rule $A \rightarrow B$, i.e., $conf(A \rightarrow B) = \dfrac{sup(A \cup B)}{sup(B)}$.

Example 4.14 Consider the frequent pattern $\{Bread, Jam, Butter\}$ in Table 3.1. An association rule that can be generated from this pattern is $\{Bread, Jam\} \rightarrow \{Butter\}$. The $confidence$ of this rule, i.e., $conf(\{Bread, Jam\} \rightarrow \{Butter\}) = \dfrac{sup(\{Bread, Jam, Butter\})}{sup(\{Bread, Jam\})} = \dfrac{2}{2} = 1$ (= 100%). If the user-specified $minConf = 0.75$ (= 75%), then $\{Bread, Jam\} \rightarrow \{Butter\}$ is said to be an interesting association rule mining. This rule says that 100% of the time, the customers purchase $Butter$ whenever they purchase $Bread$ and Jam.

The Python code to find interesting association rules from a set of frequent patterns is provided below.

Program 10: Finding Interesting Association Rules

```
from PAMI.AssociationRules.basic import confidence as alg

obj = alg.confidence('frequentPatterns.txt', minConf=0.75)
obj.mine()
obj.printResults()
obj.save("associationRulesconfidence.csv")
```

4.6 Conclusion

In this chapter, we discovered interesting patterns in transactional databases, focusing on frequent patterns and its variants. Frequent patterns reveal user behaviors and preferences, forming the basis for practical data mining. We examined key algorithms such as Apriori, ECLAT, and FP-growth and their efficiency in handling large datasets.

We also addressed the challenge of pattern overload by exploring variants like closed, maximal, and top-*k* frequent patterns. We discussed solutions to the rare item problem, such as multiple minimum supports, along with techniques for identifying correlated and relative frequent patterns.

Finally, we covered fault tolerance in pattern mining, emphasizing the need for robust patterns in the presence of data imperfections. Overall, the discussed methods and algorithms offer a solid foundation for uncovering valuable insights from transactional data.

References

1. Charu C. Aggarwal, Yan Li, Jianyong Wang, and Jing Wang. 2009. Frequent pattern mining with uncertain data. In Proceedings of the 15th ACM SIGKDD international conference on Knowledge discovery and data mining (KDD '09). Association for Computing Machinery, New York, NY, USA, 29–38.
2. Guizhen Yang. 2004. The complexity of mining maximal frequent itemsets and maximal frequent patterns. In Proceedings of the tenth ACM SIGKDD international conference on Knowledge discovery and data mining (KDD '04). Association for Computing Machinery, New York, NY, USA, 344–353.
3. Krishna Gade, Jianyong Wang, and George Karypis. 2004. Efficient closed pattern mining in the presence of tough block constraints. In Proceedings of the tenth ACM SIGKDD international conference on Knowledge discovery and data mining (KDD '04). Association for Computing Machinery, New York, NY, USA, 138–147.
4. Bing Liu, Wynne Hsu, and Yiming Ma. 1999. Mining association rules with multiple minimum supports. In Proceedings of the fifth ACM SIGKDD international conference on Knowledge discovery and data mining (KDD '99). Association for Computing Machinery, New York, NY, USA, 337–341.
5. Abdus Salam and M. Sikandar Hayat Khayal. 2012. Mining top-K frequent patterns without minimum support threshold. Knowl. Inf. Syst. 30, 1 (January 2012), 57–86.
6. Kim, W.Y., Lee, Y.K., Han, J. (2004). CCMine: efficient mining of confidence-closed correlated patterns. In: *PAKDD* (pp. 569–579).
7. Uday Kiran Rage and Masaru Kitsuregawa. 2015. Efficient discovery of correlated patterns using multiple minimum all-confidence thresholds. J. Intell. Inf. Syst. 45, 3 (December 2015), 357–377.
8. R. Uday Kiran and Masaru Kitsuregawa. 2012. Towards efficient discovery of frequent patterns with relative support. In Proceedings of the 18th International Conference on Management of Data (COMAD '12). Computer Society of India, Mumbai, Maharashtra, IND, 92–99.
9. Jhih-Jie Zeng, Guanling Lee, and Chung-Chi Lee. 2008. Mining fault-tolerant frequent patterns efficiently with powerful pruning. In Proceedings of the 2008 ACM symposium on Applied computing (SAC '08). Association for Computing Machinery, New York, NY, USA, 927–931.

10. Mohammed J. Zaki. 2000. Scalable Algorithms for Association Mining. IEEE Trans. on Knowl. and Data Eng. 12, 3 (May 2000), 372–390.
11. Mohammed J. Zaki and Karam Gouda. 2003. Fast vertical mining using diffsets. In Proceedings of the ninth ACM SIGKDD international conference on Knowledge discovery and data mining (KDD '03). Association for Computing Machinery, New York, NY, USA, 326–335.
12. Jiawei Han, Jian Pei, and Yiwen Yin. 2000. Mining frequent patterns without candidate generation. SIGMOD Rec. 29, 2 (June 2000), 1–12.

Chapter 5
Temporal Databases: Representation, Creation, and Statistics

Abstract This chapter provides a comprehensive overview of handling temporal databases using the PAMI package. Temporal databases, characterized by their time-ordered transactions, are essential for capturing and analyzing time-based data across domains such as sensor networks, satellite monitoring, and social media. We introduce the structure and representation of temporal databases, distinguishing between nonuniform and uniform types. The chapter covers practical aspects of working with these databases, including creating synthetic temporal databases for testing and converting dataframes into temporal databases. Additionally, we explore how to derive statistical details about temporal databases to understand their properties and facilitate data analysis. The techniques and tools discussed provide a solid foundation for managing, analyzing, and extracting insights from time-ordered transactional data.

5.1 Introduction

A structured certain binary temporal database, or simply a temporal database, is an organized collection of transactions ordered by time. Each transaction in this database is uniquely identified and timestamped, providing a chronological sequence of events or interactions. In particular, a transaction in a temporal database includes a transaction identifier, a relative timestamp,[1] and a set of items, typically represented in binary form to indicate their presence or absence in the data.

Temporal databases are prevalent in various real-world scenarios where time and sequence are essential. In sensor networks, each transaction might record the binary states of sensors at specific time intervals, capturing dynamic environmental changes. In satellite data, temporal databases can track the presence of certain phenomena over time, enabling detailed temporal analysis of environmental changes. Social networks also utilize temporal databases to record user interactions and activities over time, helping to uncover trends and patterns in user behavior.

[1] The timestamp of the first transaction must always start with 1.

U. K. Rage, *Hands-on Pattern Mining*,
https://doi.org/10.1007/978-981-96-6791-8_5

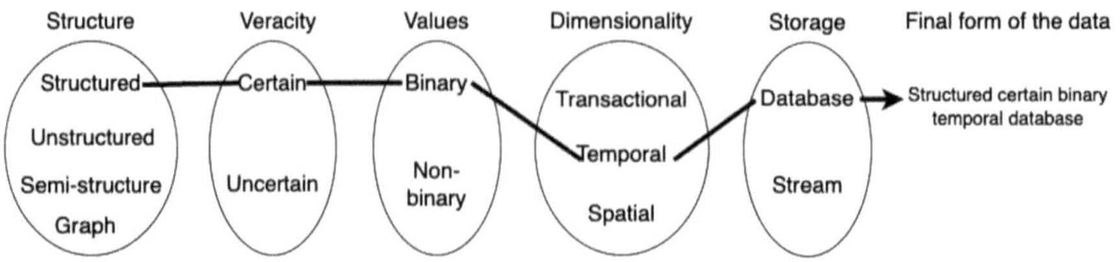

Fig. 5.1 Illustration of factors contributing to the creation of a temporal database

Figure 5.1 illustrates the complex interconnections resulting in the generation of a temporal database. This visualization aids in understanding the intricate relationships and interactions that underpin the organization and analysis of temporal data.

Temporal databases can be categorized into two types based on the interval occurrences of transactions:

- **Nonuniform Temporal Databases**: Transactions occur irregularly, with varying time intervals between successive transactions. This type is common in unpredictable scenarios or those dependent on external factors.
- **Uniform Temporal Databases [1–3]**: Transactions occur at regular, fixed intervals. A transactional database is a specific type of uniform temporal database where transactions are recorded at uniform time steps.

> **Important**

A transactional database typically represents a uniform temporal database.

Since nonuniform temporal databases cover a broader range of scenarios, the techniques developed for them can generally be applied to uniform temporal databases. Therefore, this chapter focuses primarily on nonuniform temporal databases to discuss pattern mining techniques applicable to various temporal data types. The chapter covers:

1. **Theoretical Representation**: The formal definition of a temporal database using set theory
2. **Practical Representation**: How computer systems implement and store temporal databases
3. **Synthetic Database Creation**: Methods for generating synthetic temporal databases for testing and benchmarking
4. **Dataframe Conversion**: Techniques for converting structured dataframes into temporal databases for broader data analysis
5. **Database Statistics**: Methods for deriving statistical details about a temporal database

5.2 Theoretical Representation

A temporal database consists of transactions ordered by time. Each transaction includes a transaction identifier (*tid*), a timestamp (*ts*), and a set of items. Formally:

Let $I = \{\text{item}_1, \text{item}_2, \ldots, \text{item}_n\}$, where $n \geq 1$, represent the set of all possible items. An itemset, or **pattern**, is defined as $Y = \{\text{item}_1, \text{item}_2, \ldots, \text{item}_k\} \subseteq I$, where $1 \leq k \leq n$. This subset Y represents a specific combination of items that can occur together in a transaction. A transaction is denoted as $\text{tran} = \{\text{tid}, \text{ts}, Y\}$, where $\text{tid} \in \mathbb{R}^+$ is the *transaction identifier*, a unique number for each transaction. The tid ensures distinct transaction identification. The $\text{ts} \geq 1$ represents the relative *timestamp* of a transaction. Multiple transactions can have the same timestamp. The set $Y \subseteq I$ includes the items present in this transaction. A temporal database, denoted as $TempDB$, is a collection of such transactions, formally defined as $TempDB = \{\text{tran}_1, \text{tran}_2, \ldots, \text{tran}_m\}$, where $m \geq 1$ represents the number of transactions.

Example 5.1 Consider the set of items $I = \{\text{Bread}, \text{Jam}, \text{Butter}, \text{Book}, \text{Pen}\}$ available in a supermarket. Table 5.1a and b present the horizontal and vertical formats of a temporal database, respectively. This database is based on the irregular purchases made by five anonymous customers. For simplicity, the concepts will be explained using the horizontal format shown in Table 5.1a.

In the first transaction, $\text{tran}_1 = \{1 : 1 : \text{Bread}, \text{Jam}, \text{Butter}\}$, the number 1 represents the transaction identifier (or *tid*), 1 is the relative timestamp (or *ts*), and {Bread, Jam, Butter} represent the items purchased in that transaction. This transaction indicates that the first customer has purchased the items "Bread," "Jam," and "Butter" at the timestamp equal to 1.

Note that multiple transactions can share the same timestamp, and missing timestamps are possible. This reflects the irregular nature of nonuniform temporal databases, where events do not occur at regular intervals.

Table 5.1 Hypothetical temporal database of a supermarket

(a) Horizontal format

tid	*ts*	Items
1	1	Bread, Jam, Butter
2	3	Bread, Book, Pen
3	3	Jam, Butter
4	5	Bread, Jam, Butter, Pen
5	8	Book, Pen

(b) Vertical format

tid	*ts*	Bread	Jam	Butter	Book	Pen
1	1	1	1	1	0	0
2	3	1	0	0	1	1
3	3	0	1	1	0	0
4	5	1	1	1	0	1
5	8	0	0	0	1	1

> **Key Properties of a Temporal Database**

- Multiple transactions can share a common timestamp.
- Timestamps need not be continuous in the data.

5.3 Practical Representation

Temporal databases are typically stored as files. To ensure proper creation and management of these files, follow these rules:

- **One Transaction per Line**: Each line represents a single transaction. The line number implicitly serves as the transaction identifier, so it is not explicitly stored in the file. Only the timestamp and items are recorded.
- **Relative Timestamp**: Each transaction must have a relative timestamp, starting with 1. Convert absolute timestamps, such as "2024-01-01 00:00:00," into relative timestamps if necessary.
- **Unique Items per Transaction**: Items must appear only once per line and can be listed in any order.
- **Positioning of Timestamp and Items**: Each transaction begins with a timestamp followed by items. Do not create transactions with only a timestamp and no items.
- **Delimiter Separation**: Use a delimiter, such as a space or tab, to separate elements. The default delimiter in PAMI algorithms is the `tab`, but other delimiters like commas or spaces can also be used.

The format of a transaction in a temporal database is

$$timestamp\langle sep\rangle item_1\langle sep\rangle item_2\langle sep\rangle item_3\langle sep\rangle \cdots$$

Example 5.2 With a tab delimiter, the temporal database in Table 5.1 would look like this:

```
1    Bread    Jam Butter
3    Bread    Book    Pen
3    Jam Butter
5    Bread    Jam Butter  Pen
8    Book     Pen
```

! Attention

Do not create a temporal database with transactions containing only timestamps.

> **Important**
>
> Tab is the default separator to distinguish the timestamp and items in a line.

5.4 Creating Synthetic Temporal Databases

The PAMI package provides a versatile approach for generating different types of synthetic temporal databases, which are essential for testing and developing algorithms in data mining. Users can customize various parameters, such as the number of transactions, the total number of items, average transaction length, probability of multiple transactions sharing the same timestamp, and the probability of skipping transactions at subsequent timestamps.

The following code snippet demonstrates how to generate a synthetic temporal database with 100,000 transactions, each containing an average of 10 items from a set of 1,000 possible items:

Program 1: Generating Synthetic Temporal Database

```
from PAMI.extras.syntheticDataGenerator import TemporalDatabase as db

obj = db.TemporalDatabase(databaseSize=100000, avgItemsPerTransaction=10, numItems=1000, occurrenceProbabilityOfSameTimestamp=0, occurrenceProbabilityToSkipSubsequentTimestamp=0, sep='\t')
obj.create()
obj.save('temporalDatabase.csv')
#read the generated transactions into a dataframe
temporalDataFrame=obj.getTransactions()
#stats
print('Runtime: ' + str(obj.getRuntime()))
print('Memory (RSS): ' + str(obj.getMemoryRSS()))
print('Memory (USS): ' + str(obj.getMemoryUSS()))
```

5.5 Deriving a Temporal Database from a Dataframe

PAMI also allows converting a dataframe into a temporal database, which is ideal for transaction-based data analysis. Below is a Python code snippet showing how to perform this conversion:

Program 2: Converting a Dataframe into a Temporal Database

```
from PAMI.extras.convert import DF2DB as alg
import pandas as pd
import numpy as np

#creating a 5 x 5 dataframe with random values
data = np.random.randint(1, 100, size=(5, 5))
dataFrame = pd.DataFrame(data,
             columns=['Item1', 'Item2', 'Item3', 'Item4', 'Item5']
            )
# Adding a timestamp column with specific values
timestamps = [1, 3, 3, 5, 8]
dataFrame.insert(0, 'timestamp', timestamps)

#converting the database into a temporal database by
#considering values greater than or equal to 36
obj = alg.DF2DB(dataFrame)
obj.convert2TemporalDatabase(oFile='temporalDB.csv', condition='>=', thresholdValue=36)
print('Runtime: ' + str(obj.getRuntime()))
print('Memory (RSS): ' + str(obj.getMemoryRSS()))
print('Memory (USS): ' + str(obj.getMemoryUSS()))
```

5.6 Knowing the Statistical Details

The `dbStats` sub-package in PAMI's `extras` module provides detailed statistical information about a temporal database. This functionality is crucial for understanding the properties and characteristics of the database. The statistical details include:

1. Database size
2. Total number of items in a database
3. Minimum, average, and maximum lengths of the transactions
4. Standard deviation of transactional sizes
5. Variance in transaction sizes
6. Sparsity
7. Frequencies of the items
8. Distribution of transactional lengths
9. Minimum, average, and maximum inter-arrival time of the transactions

10. Minimum, average, and maximum periodicity of the items

Here is an example of how to use the dbStats to obtain these statistics:

Program 3: Deriving the Statistical Details

```
from PAMI.extras.dbStats import TemporalDatabase as stat

obj = stat.TemporalDatabase("temporalDatabase.csv")
obj.run()
obj.printStats()
obj.plotGraphs()
```

5.7 Conclusion

This chapter explored the essential aspects of working with temporal databases using the PAMI package. We began by understanding the structure of temporal databases, which record transactions in a time-ordered manner and are used in real-world applications like sensor networks, satellite data, and social networks.

We discussed temporal databases' theoretical and practical representations, highlighting the differences between nonuniform and uniform temporal databases and their respective uses. This chapter also covered the methods for creating synthetic temporal databases, demonstrating how to generate and customize databases for testing and algorithm development.

Additionally, we illustrated how to convert a dataframe into a temporal database, providing a practical approach to data analysis. Lastly, we highlighted the importance of statistical details in understanding the properties of temporal databases, showing how to derive and interpret these statistics to analyze and utilize the data better.

References

1. Jiawei Han, Wan Gong, and Yiwen Yin. 1998. Mining segment-wise periodic patterns in time-related databases. In Proceedings of the Fourth International Conference on Knowledge Discovery and Data Mining (KDD'98). AAAI Press, 214–218.
2. Jiong Yang, Wei Wang, and Philip S. Yu. 2003. Mining Asynchronous Periodic Patterns in Time Series Data. IEEE Trans. on Knowl. and Data Eng. 15, 3 (March 2003), 613–628.
3. Syed Khairuzzaman Tanbeer, Chowdhury Farhan Ahmed, Byeong-Soo Jeong, and Young-Koo Lee. 2009. Discovering Periodic-Frequent Patterns in Transactional Databases. In Proceedings of the 13th Pacific-Asia Conference on Advances in Knowledge Discovery and Data Mining (PAKDD '09). Springer-Verlag, Berlin, Heidelberg, 242–253.

Chapter 6
Pattern Discovery in Temporal Databases

Abstract Periodic-frequent pattern mining is a critical technique for analyzing temporal data to identify recurring trends and patterns. However, traditional models face significant challenges, such as the rare item problem, where uniform frequency and periodicity assumptions can lead to either the exclusion of patterns involving rare items or the generation of excessive, trivial patterns. Additionally, these models often fail to capture patterns with partial periodicity, limiting their applicability in real-world scenarios where periodic behavior may be intermittent. To address these issues, advancements such as periodic-correlated pattern mining have been developed, incorporating measures like all-confidence and periodic-all-confidence to balance the significance of frequent and rare items. Furthermore, partial periodic pattern discovery models relax strict periodicity constraints, allowing for identifying patterns with intermittent periodic behavior. These innovations enhance the ability to extract valuable insights from complex temporal datasets, improving decision-making and strategic planning.

6.1 Introduction

In the preceding chapter, we discussed temporal databases' construction, practical representation, and statistical analysis. Building on this foundation, this chapter focuses on the analytical aspects of temporal data, specifically extracting and examining meaningful patterns. One key area of interest is the identification of periodic-frequent patterns regularities that recur frequently at consistent intervals. These patterns are crucial for understanding data trends and behaviors over time.

Periodic-frequent patterns can be categorized into two main types based on their occurrence behavior: *perfect periodic-frequent patterns* and *partial periodic-frequent patterns*. A perfect periodic-frequent pattern appears with high frequency throughout the dataset and exhibits regular intervals consistently across the entire timeframe. In contrast, a partial periodic-frequent pattern occurs frequently but shows periodic behavior only during specific data segments. Due to real-world complexities and inherent noise, partial periodic-frequent patterns are often more insightful, offering a nuanced view of periodic behaviors that may not be uniform

U. K. Rage, *Hands-on Pattern Mining*,
https://doi.org/10.1007/978-981-96-6791-8_6

throughout the dataset. This chapter focuses on perfect and partial periodic-frequent patterns, providing methods for users to choose the most suitable pattern mining technique based on their application requirements.

This chapter delves into several critical areas of mining temporal databases to uncover these patterns:

1. **Periodic-Frequent Pattern Discovery**: We will define a (perfect) periodic-frequent pattern, explore the search space involved in identifying these patterns, discuss the Apriori property, and outline various algorithms for discovering such patterns.
2. **Handling Redundancy Problem in Periodic-Frequent Patterns**: This section addresses the issue of redundancy in pattern mining. Techniques such as mining closed periodic-frequent patterns, identifying maximal periodic-frequent patterns, and selecting top-*k* periodic-frequent patterns will be discussed to streamline pattern discovery and reduce redundant findings.
3. **Rare Item Problem and Solutions**: We will examine the challenges of mining infrequent or rare items within temporal data. Solutions and strategies for addressing the rare item problem will be presented to improve the comprehensiveness of pattern mining.
4. **Finding Partial Periodic Patterns**: Different patterns exhibiting partial periodic behavior will be explored. This includes methods for discovering patterns that show periodic behavior only in specific periods, offering practical insights into the variability of periodic trends.

6.2 Periodic-Frequent Patterns

6.2.1 The Basic Model

Chapter 5 introduced the foundational concepts of temporal databases, including key terms such as "pattern," "transaction," and "temporal database." For clarity and consistency, we will use these terms throughout this chapter. We recommend reviewing Sect. 5.2 for readers unfamiliar with these concepts.

Definition 6.1 (Temporal Occurrences of a Pattern) Let TS denote the set of all timestamps in $TempDB$. Let $P \subseteq I$ be a pattern. If $P \subseteq Y$, we say P occurs in Y (or Y contains P). Let $ts_i^P \in TS, i \geq 1$, denote the occurrence timestamp of pattern P in a transaction. Let $TS^P \subseteq TS$ denote the set of all timestamps containing P in $TempDB$.

Example 6.1 Consider the pattern $\{Jam, Butter\}$ in Table 6.1. This pattern initially occurs in the first transaction, whose timestamp is 1. Thus, $ts_1^{\{Jam,Butter\}} = 1$. Similarly, $ts_2^{\{Jam,Butter\}} = 3$ and $ts_3^{\{Jam,Butter\}} = 5$. The set of all timestamps containing the pattern $\{Jam, Butter\}$, i.e., $TS^{\{Jam,Butter\}} = \{1, 3, 5\}$. In other words, the items "Jam" and "Butter" were co-purchased by customers at timestamps 1, 3, and 5.

Table 6.1 Temporal database

tid	ts	Items
1	1	Bread, Jam, Butter
2	3	Bread, Book, Pen
3	3	Jam, Butter
4	5	Bread, Jam, Butter, Pen
5	8	Butter, Book, Pen

Definition 6.2 (Support of a Pattern) The *support* of a pattern P, denoted as $sup(P)$, is defined as $|TS^P|$. (Support can also be expressed as a percentage of the database size.)

Example 6.2 The *support* of the pattern $\{Jam, Butter\}$ in Table 6.1 is given by $sup(\{Jam, Butter\}) = |TS^{\{Jam, Butter\}}| = |\{1, 3, 5\}| = 3$ or $60\%(= \frac{3\times 100}{5})$.

Definition 6.3 (Frequent Pattern) A pattern P is said to be *frequent* if $sup(P) \geq minSup$, where $minSup$ represents the user-specified minimum support value.

Example 6.3 If the user-specified $minSup = 2$, then $\{Jam, Butter\}$ is a frequent pattern because $sup(\{Jam, Butter\}) \geq minSup$.

Definition 6.4 (Inter-Arrival Times of a Pattern) Let ts_a^P and ts_b^P (with $a < b$) represent two consecutive occurrences of the pattern P in $TempDB$. An inter-arrival time of P, denoted as iat_k^P, is defined as $ts_b^P - ts_a^P$. The set of all inter-arrival times of P is denoted as $IAT^P = \{iat_1^P, iat_2^P, \cdots, iat_x^P\}$, where $x = |TS^P| - 1$.

Example 6.4 Consider the pattern $\{Jam, Butter\}$, which appears at timestamps 1, 3, and 5. The first inter-arrival time for this pattern, i.e., $iat_1^{\{Jam, Butter\}} = 3 - 1 = 2$. Similarly, the second inter-arrival time of this pattern, i.e., $iat_2^{\{Jam, Butter\}} = 5 - 3 = 2$. The set of all inter-arrival times of $\{Jam, Butter\}$, i.e., $IAT^{\{Jam, Butter\}} = \{2, 2\}$.

Definition 6.5 (Periodicity of a Pattern) Let $ts_{ini} = 0$ and $ts_{fin} = \max(ts_i \mid \forall ts_i \in TS)$ be the initial and final timestamps of the database, respectively. The time consumed for the initial appearance of P in the temporal database $TempDB$ is $iat_{consumed}^P = (ts_1^P - ts_{ini})$. The time elapsed after the final appearance of P in $TempDB$ is $iat_{elapsed}^P = (ts_{fin} - \max(ts_k^P \mid \forall ts_k^P \in TS^P))$. The *periodicity* of P in $TempDB$, denoted as $per(P)$, is defined as $\max(iat_q^P \mid \forall \{IAT^P \cup iat_{consumed}^P \cup iat_{elapsed}^P\})$.

Example 6.5 For the temporal database shown in Table 6.1, the initial and final timestamps are 0 and 8, respectively. Thus, $ts_{ini} = 0$ and $ts_{fin} = 8$. The time taken for the initial occurrence of the pattern $\{Jam, Butter\}$ is $iat_{consumed}^{\{Jam, Butter\}} = (ts_1^{\{Jam, Butter\}} - ts_{ini}) = 1 - 0 = 1$. The time elapsed after the final occurrence of $\{Jam, Butter\}$ is $iat_{elapsed}^{\{Jam, Butter\}} = (ts_{fin} - \max(TS^{\{Jam, Butter\}})) = 8 - 5 = 3$. The *periodicity* of $\{Jam, Butter\}$ is $per(\{Jam, Butter\}) = \max(IAT^{\{Jam, Butter\}} \cup iat_{consumed}^{\{Jam, Butter\}} \cup iat_{elapsed}^{\{Jam, Butter\}}) = \max(2, 2, 1, 3) = 3$.

Definition 6.6 (Periodic-Frequent Pattern) A frequent pattern P is considered a *periodic-frequent pattern* if $per(P) \leq maxPrd$, where $maxPrd$ is the user-specified *maximum periodicity* threshold value.

Example 6.6 If the user-specified $maxPrd = 3$, the frequent pattern $\{Jam, Butter\}$ is a periodic-frequent pattern because $per(\{Jam, Butter\}) \leq maxPrd$.

Definition 6.7 (Problem Definition) Given a temporal database ($TempDB$) and the user-specified *minimum support* ($minSup$) and *maximum periodicity* ($maxPrd$) values, the problem is to discover all periodic-frequent patterns in TDB that have support no less than $minSup$ and periodicity no more than $maxPrd$.

Note The *inter-arrival times* and *periodicity* of a pattern can also be expressed as percentages of ts_{fin}. However, the *periodicity* is expressed in counts throughout this book for brevity.

6.2.2 *Search Space and Apriori Property*

In periodic-frequent pattern mining, the goal is to identify frequent patterns in a dataset that follow a regular, repeating interval. To achieve this, it is necessary to search through many potential patterns. The search space and the Apriori property are crucial concepts in managing this process.

6.2.2.1 Search Space

1. **Itemset Lattice**: The itemset lattice represents the search space for periodic-frequent pattern mining. An itemset lattice includes all possible combinations of items from the dataset. For a database with $|I|$ items, the itemset lattice contains all possible itemsets of size one up to size $|I|$.
2. **Size of Search Space**: The total number of possible itemsets is $2^{|I|} - 1$. This is because each item can either be included in a pattern or not, leading to $2^{|I|}$ possible combinations. We subtract 1 to exclude the empty set. For instance, if there are five items in the database, there are $2^5 - 1 = 31$ possible non-empty itemsets. (See Chap. 4 for more information.)
3. **Challenge**: Directly searching through all these itemsets is computationally infeasible, especially as the number of items increases. The number of potential patterns grows exponentially with the number of items, necessitating strategies to efficiently reduce the search space.

6.2.2.2 Apriori Property

1. **Definition**: The Apriori property is a fundamental principle used to reduce the search space in frequent pattern mining. It states that "all non-empty subsets of a periodic-frequent pattern must also be periodic-frequent patterns."

2. **Implication**: If a pattern is identified as a periodic-frequent pattern, then every subset of this pattern must also be periodic frequent. For example, if the pattern $\{A, B, C\}$ is found to be periodic frequent, then the patterns $\{A, B\}$, $\{A, C\}$, and $\{B, C\}$ must also be periodic frequent.
3. **Utility**: We can significantly reduce the number of candidate patterns to evaluate by applying this property. Instead of checking all possible patterns, we focus only on supersets of known periodic-frequent patterns. This reduction in the search space is because if a large itemset is periodic frequent, all of its smaller subsets must be periodic frequent. Conversely, if a subset is not periodic frequent, any larger itemset containing it cannot be periodic frequent either.
4. **Example**: Suppose we are searching for patterns in a dataset of five items and have identified that $\{A, B, C\}$ as a periodic-frequent pattern. According to the Apriori property, any sub-pattern that includes $\{A, B, C\}$, such as $\{A, B, C\}$, must also be periodic frequent if it meets the support and periodicity criteria. This allows us to avoid evaluating larger itemsets that do not contain periodic-frequent subsets, focusing our efforts on more promising candidates.

In summary, the Apriori property is a critical tool for efficiently mining periodic-frequent patterns. By leveraging this property, we can reduce the vast search space and make identifying meaningful patterns more manageable.

6.2.3 *Finding Periodic-Frequent Patterns*

Several algorithms have been proposed in the literature for finding periodic-frequent patterns, including PFP-growth [1], PFP-growth++ [2], and PF-ECLAT [3]. While there is no universally accepted best algorithm for finding periodic-frequent patterns across all temporal databases, the PFP-growth++ algorithm is often preferred due to its generally faster performance than other algorithms. Below is an example Python script demonstrating how to find periodic-frequent patterns using the PFP-growth++ algorithm, which is available in the PAMI package.

Program 1: Finding Periodic-Frequent Patterns

```python
from PAMI.periodicFrequentPattern.basic import PFPGrowthPlus as
↪ alg # Import the algorithm

obj = alg.PFPGrowthPlus(iFile='Temporal_T10I4D100K.csv',
↪ minSup=100, maxPer='2000', sep='\t') # Initialize
obj.mine()
obj.save('periodicFrequentPatterns.txt')

patternsDF = obj.getPatternsAsDataFrame()
```

```
print('Patterns: ' + str(len(patternsDF)))
print('Runtime: ' + str(obj.getRuntime()))
print('Memory (RSS): ' + str(obj.getMemoryRSS()))
print('Memory (USS): ' + str(obj.getMemoryUSS()))
```

6.3 Popular Variants of Periodic-Frequent Patterns

The primary goal of the basic periodic-frequent pattern model is to identify all patterns that meet the user-specified *minSup* and *maxPrd* criteria in a temporal database. However, this approach can generate many patterns, which may be redundant or of limited interest depending on the user's requirements or specific application needs.

Example 6.7 The basic periodic-frequent pattern model not only identifies $\{Jam, Butter\}$ as a periodic-frequent pattern in Table 3.1a but also considers all of its non-empty subsets, i.e., $\{Jam\}$ and $\{Butter\}$, as periodic-frequent patterns. Due to this redundancy, users might find these subsets, $\{Jam\}$ and $\{Butter\}$, less attractive.

To address this issue, researchers have developed methods to find *closed periodic-frequent patterns* [4], *maximal periodic-frequent patterns* [5], and *top-k rperiodic-frequent patterns* [6]. This section will briefly explore these variants and discuss the methods for identifying them.

6.3.1 *Closed Periodic-Frequent Patterns*

A periodic-frequent pattern is considered a closed periodic-frequent pattern if none of its supersets share the same support and periodicity. Let PFP and $CPFP$ denote the sets of periodic-frequent and closed periodic-frequent patterns, respectively, generated from a temporal database with given values of $minSup$ and $maxPrd$. The relationship between these sets is $CPFP \subseteq PFP$ (or equivalently, $|CPFP| \leq |PFP|$). In other words, closed periodic-frequent patterns are smaller in number than periodic-frequent patterns. More importantly, closed periodic-frequent patterns provide a lossless representation of periodic-frequent patterns, meaning that the complete set of periodic-frequent patterns can be reconstructed from the closed periodic-frequent patterns without losing any information.

Example 6.8 Consider the following periodic-frequent patterns in Table 6.1: $\{Jam\}$, $\{Butter\}$, and $\{Jam, Butter\}$. The relationships among these patterns are as follows: $\{Jam\}$ and $\{Butter\}$ are subsets of $\{Jam, Butter\}$. The support values for these patterns are: $sup(\{Jam\}) = 3$, $sup(\{Butter\}) = 4$, and $sup(\{Jam, Butter\}) = 3$. The periodicity values are: $per(\{Jam\}) = 3$,

$per(\{Butter\}) = 3$, and $per(\{Jam, Butter\}) = 3$. Since the support and periodicity of $\{Jam\}$ are the same as those of its superset $\{Jam, Butter\}$, $\{Jam\}$ can be disregarded as it is redundant and can be derived from $\{Jam, Butter\}$ without any loss of information. However, $\{Butter\}$ and $\{Jam, Butter\}$ have different support values, so $\{Butter\}$ cannot be derived from $\{Jam, Butter\}$ without loss of information. Therefore, $\{Butter\}$ and $\{Jam, Butter\}$ are considered closed periodic-frequent patterns.

The procedure for finding closed periodic-frequent patterns in a temporal database is outlined below:

Program 2: Finding Closed Periodic-Frequent Patterns

```
from PAMI.periodicFrequentPattern.closed import CPFPMiner as
    alg

obj = alg.CPFPMiner(iFile='Temporal_T10I4D100K.csv',
    minSup=100, maxPer=2000, sep='\t')

obj.mine()
obj.save('closedPeriodicFrequentPatterns.txt')

patternsDF = obj.getPatternsAsDataFrame()
print('Patterns: ' + str(len(patternsDF)))
print('Runtime: ' + str(obj.getRuntime()))
print('Memory (RSS): ' + str(obj.getMemoryRSS()))
print('Memory (USS): ' + str(obj.getMemoryUSS()))
```

Important

Closed periodic-frequent patterns represent a lossless subset of periodic-frequent patterns.

6.3.2 *Maximal Periodic-Frequent Patterns*

A periodic-frequent pattern is considered a maximal periodic-frequent pattern if none of its supersets are periodic-frequent. Let PFP, $CPFP$, and $MPFP$ denote the sets of periodic-frequent patterns, closed periodic-frequent patterns, and maximal periodic-frequent patterns, respectively, generated from a temporal database with given values of $minSup$ and $maxPrd$. The relationship among these sets is given by $MPFP \subseteq CPFP \subseteq PFP$ (or equivalently, $|MPFP| \leq |CPFP| \leq |PFP|$). Unlike closed periodic-frequent patterns, maximal periodic-frequent patterns offer a lossy representation because they do not retain the exact support and periodicity information of all the periodic-frequent patterns.

Example 6.9 Considering the previous example, consider the closed periodic-frequent patterns $\{Butter\}$ and $\{Jam, Butter\}$. Among these, $\{Jam, Butter\}$ is a maximal periodic-frequent pattern because none of its supersets are periodic frequent. Although we can derive all of its subset periodic-frequent patterns from $\{Jam, Butter\}$, the exact $support$ and $periodicity$ values for these subsets are not determinable. Thus, maximal periodic-frequent patterns represent a lossy approximation of the complete set of frequent patterns.

> **Important**
>
> Maximal periodic-frequent patterns represent a lossy subset of the frequent patterns.

The procedure for finding maximal periodic-frequent patterns in a temporal database is outlined below:

Program 3: Finding Maximal Periodic-Frequent Patterns

```
from PAMI.periodicFrequentPattern.maximal  import MaxPFGrowth
↪   as alg

obj = alg.MaxPFGrowth(iFile='Temporal_T10I4D100K.csv',
↪   minSup=100, maxPer=2000, sep='\t')

obj.mine()
obj.save('maximalPeriodicFrequentPatterns.txt')

patternsDF = obj.getPatternsAsDataFrame()
print('Patterns: ' + str(len(patternsDF)))
print('Runtime: ' + str(obj.getRuntime()))
print('Memory (RSS): ' + str(obj.getMemoryRSS()))
print('Memory (USS): ' + str(obj.getMemoryUSS()))
```

6.3.3 Top-k Periodic-Frequent Patterns

A common challenge with traditional periodic-frequent pattern mining is determining the appropriate *minimum support* and *maximum periodicity* values for a given temporal database. To address this challenge, researchers have developed the concept of top-*k* periodic-frequent pattern mining. This approach focuses on identifying the top-*k* patterns that exhibit the lowest periodicity in the dataset, regardless of their support values. This method is beneficial when the goal is to discover the most significant patterns based on periodicity rather than predefined thresholds.

The following Python script demonstrates how to find the top-*k* periodic-frequent patterns using the PAMI package:

Program 4: Finding Top-*k* Periodic-Frequent Patterns

```
from PAMI.periodicFrequentPattern.topk.kPFPMiner import
↪  kPFPMiner as alg

obj = alg.kPFPMiner(iFile='Temporal_T10I4D100K.csv', k=1000,
↪  sep='\t')
obj.mine()

obj.save('topkPeriodicFrequentPatterns.txt')

kPatternsDF = obj.getPatternsAsDataFrame()
print('#Patterns: ' + str(len(kPatternsDF)))
print('Runtime: ' + str(obj.getRuntime()))
print('Memory (RSS): ' + str(obj.getMemoryRSS()))
print('Memory (USS): ' + str(obj.getMemoryUSS()))
```

6.4 Main Issues of Periodic-Frequent Pattern Mining

The basic model of periodic-frequent pattern mining faces two significant challenges:

1. **The Rare Item Problem:** The basic model of periodic-frequent patterns relies on a single $minSup$ and $maxPrd$ to assess the interestingness of patterns across the entire dataset. This approach implicitly assumes that all items have uniform frequencies or similar temporal occurrence behaviors, which is rarely true in real-world applications. In many scenarios, some items appear frequently, while

others occur infrequently. This variation in occurrence behavior leads to two potential problems:

- Setting a high *minSup* or a low *maxPrd* value may result in missing periodic-frequent patterns that contain rare items, as these items often do not meet the specified constraints.
- To capture patterns involving frequent and rare items, one might need to set a low *minSup* and a high *maxPrd*. However, this can lead to a combinatorial explosion, generating an overwhelming number of patterns, many of which may be uninteresting or irrelevant to the user or application.

This dilemma is commonly referred to as the "rare item problem."

2. **Inability to Find Partially Periodically Occurring Patterns:** The basic periodic-frequent pattern model enforces a strict requirement that all inter-arrival times of a pattern must be within the user-specified *maxPrd* threshold. This rigid criterion can cause the model to overlook interesting patterns that exhibit partial periodic behavior in the data, thereby missing potentially valuable insights.

In the following sections, we will explore various approaches described in the literature to address these two issues.

6.5 Addressing the Rare Item Problem

The rare item problem, a significant challenge in periodic-frequent pattern mining, arises due to the inherent assumption that all items in a dataset exhibit similar frequencies and temporal behaviors. This assumption is often violated in real-world datasets, where some items appear frequently, while others occur only sporadically. To address this problem, researchers have introduced the concept of *periodic-correlated pattern mining*. This approach extends the basic periodic-frequent pattern model by incorporating additional constraints that account for both the frequency and the correlation of items within a pattern, allowing for the identification of patterns involving rare items without overwhelming the user with trivial results.

6.5.1 Periodic-Correlated Pattern Mining

Periodic-correlated pattern mining [7] utilises the concept of *correlation* between items within a pattern, alongside the traditional support and periodicity constraints. This method ensures that patterns containing rare items are not overlooked due to the global application of a single minimum support and maximum periodicity threshold values on the entire dataset. In particular, this model does so by incorporating an *all-confidence* measure, which balances the influence of frequent and rare items within the same pattern.

Definition 6.8 (Periodic-Correlated Pattern) A pattern P is considered a periodic-correlated pattern if it satisfies the following constraints:

$$sup(P) \geq minSup \tag{6.1}$$

$$allConf(P) \geq minAllConf \tag{6.2}$$

$$per(P) \leq maxPer \tag{6.3}$$

$$PC(P) \leq maxPeriodicAllConf. \tag{6.4}$$

Here, the different terms are defined as follows:

- $sup(P)$: The support of the pattern P, which must be greater than or equal to a user-specified minimum support threshold, $minSup$.
- $allConf(P)$: The all-confidence measure, defined as:

$$allConf(P) = \frac{sup(P)}{\max(sup(i_j) \mid \forall i_j \in P)},$$

  which must be greater than or equal to a minimum all-confidence threshold, $minAllConf$. This measure accounts for the balance between the frequent and rare items within the pattern.
- $per(P)$: The periodicity of the pattern P, which must be less than or equal to a user-specified maximum periodicity threshold, $maxPer$.
- $PC(P)$: The periodic-all-confidence measure, defined as:

$$PC(P) = \frac{\mid \widehat{IAT^P} \mid}{\min(sup(i_j) \mid \forall i_j \in P)} - 1,$$

  which must be less than or equal to a maximum periodic-all-confidence threshold, $maxPeriodicAllConf$. Here, $\widehat{IAT^P} \subseteq IAT^P$ represents the set of inter-arrival times that are less than a user-specified *maximum inter-arrival time*.

The constraints on the all-confidence measure and the periodic-all-confidence measure ensure that the pattern is not only frequent and periodic but also that it reflects a meaningful correlation between its constituent items. This approach allows for the inclusion of patterns that involve rare items without being overwhelmed by trivial or irrelevant patterns.

6.5.2 Implementation Example: Finding Periodic-Correlated Patterns

The following Python code illustrates how periodic-correlated patterns can be identified within a temporal database. It uses the PAMI package to find such patterns. This example employs the EPCPGrowth algorithm, which is designed to mine periodic-correlated patterns by considering both the frequency and periodicity of items and their correlation.

Program 5: Finding Periodic-Correlated Patterns

```
from PAMI.periodicCorrelatedPattern.basic import EPCPGrowth as
↪ alg

# Initialize the EPCPGrowth algorithm with the appropriate
↪ parameters
obj = alg.EPCPGrowth(
        iFile='Temporal_T10I4D100K.csv',
        minSup=100,
        minAllConf=0.7,
        maxPer=2000,
        maxPerAllConf=1.5,
        sep='\t')

# Mine the periodic-correlated patterns
obj.mine()

# Save the patterns to a file
obj.save('correlatedPeriodicFrequentPatterns.txt')

# Retrieve the patterns as a DataFrame
correlatedPFPs = obj.getPatternsAsDataFrame()

# Display summary information
print('#Patterns: ' + str(len(correlatedPFPs)))
print('Runtime: ' + str(obj.getRuntime()))
print('Memory (RSS): ' + str(obj.getMemoryRSS()))
print('Memory (USS): ' + str(obj.getMemoryUSS()))
```

6.6 Finding Partial Periodic Patterns

In many real-world applications, certain patterns may not exhibit consistent periodic behavior throughout a temporal database. Instead, these patterns might show periodicity only during certain intervals or under specific conditions. To identify such patterns, researchers have developed various models for mining partial periodic patterns. This section explores three prominent models for discovering interesting patterns that exhibit partial periodic behavior.

6.6.1 Partial Periodic-Frequent Patterns

Partial periodic-frequent patterns [8] are a generalization of periodic-frequent patterns. They relax the strict requirement that a pattern must consistently occur within a specified period throughout the entire database. Instead, they allow patterns to be identified as periodic frequent even if they only exhibit periodic behavior for a portion of the time.

Definition 6.9 (Periodic Ratio of a Pattern) The *periodic ratio* of a pattern P, denoted as $PR(P)$, quantifies the proportion of P's occurrences in the database that are periodic. It is defined as follows:

$$PR(P) = \frac{|\widehat{IAT^P}|}{sup(P) - 1}, \tag{6.5}$$

where $sup(P)$ is the support of pattern P, and $|\widehat{IAT^P}|$ is the number of inter-arrival times within the user-specified maximum periodicity threshold.

Definition 6.10 (Partial Periodic-Frequent Pattern) A pattern P is considered a *partial periodic-frequent pattern* if it satisfies the following conditions:

$$sup(P) \geq minSup \tag{6.6}$$

$$per(P) \leq maxPer \tag{6.7}$$

$$PR(P) \geq minPR, \tag{6.8}$$

where $minSup$ is the minimum support threshold, $maxPer$ is the maximum periodicity threshold, and $minPR \in (0, 1)$ is the user-specified minimum periodic ratio. The minimum periodic ratio ensures that a pattern is considered interesting only if it maintains a certain level of periodic occurrences in the database.

To find all partial periodic-frequent patterns in a temporal database, you can use the following Python code. This code utilizes the `GPFgrowth` algorithm from the PAMI package to mine patterns that meet the specified support, periodicity, and periodic ratio constraints.

Program 6: Finding Partial Periodic-Frequent Patterns

```python
from PAMI.partialPeriodicFrequentPattern.basic \
    import GPFgrowth as alg

# Initialize the GPFgrowth algorithm with the required
↪   parameters
obj = alg.GPFgrowth(
```

```
        iFile='Temporal_T10I4D100K.csv',
        minSup=100,
        maxPer=2000,
        minPR=0.5,
        sep='\t')

# Mine the partial periodic-frequent patterns
obj.mine()

# Save the patterns to a file
obj.save('partialPeriodicFrequentPatterns.txt')

# Retrieve the patterns as a DataFrame
PPFPs = obj.getPatternsAsDataFrame()

# Display summary information
print('#Patterns: ' + str(len(PPFPs)))
print('Runtime: ' + str(obj.getRuntime()))
print('Memory (RSS): ' + str(obj.getMemoryRSS()))
print('Memory (USS): ' + str(obj.getMemoryUSS()))
```

6.6.2 Partial Periodic Patterns

In certain temporal databases, patterns may not occur with consistent periodicity throughout the dataset but may still exhibit periodic behavior over specific intervals. To capture such behavior, the concept of partial periodic patterns [9] is introduced. This approach identifies patterns that may not be frequent in the entire dataset but occur periodically within certain segments.

Definition 6.11 (Periodic Support of a Pattern) The *periodic support* of a pattern P, denoted as $PS(P)$, is the count of occurrences where P is considered periodic. An occurrence of P is considered periodic if the inter-arrival time (the time between consecutive occurrences) is within the user-specified *maximum inter-arrival time* ($maxIAT$). Formally,

$$PS(P) = |\widehat{IAT^P}|, \tag{6.9}$$

where $\widehat{IAT^P} \subseteq IAT^P$ represents the subset of inter-arrival times that are less than or equal to the user-defined $maxIAT$.

Definition 6.12 (Partial Periodic Pattern) A pattern P is defined as a *partial periodic pattern* if its periodic support $PS(P)$ meets or exceeds a user-specified threshold, known as the *minimum periodic support* ($minPS$). In other words,

$$PS(P) \geq minPS. \tag{6.10}$$

This means that P must have a sufficient number of periodic occurrences within the dataset to be considered a partial periodic pattern.

Definition 6.13 (Problem Definition) Given a temporal database $TempDB$, a *maximum inter-arrival time* $maxIAT$, and a *minimum periodic support* $minPS$, the task is to find all patterns P in $TempDB$ such that the periodic support $PS(P)$ is no less than $minPS$.

The partial periodic pattern mining search space is $2^{|I|} - 1$, where $|I|$ represents the total number of distinct items in the database. Given the vast size of this search space, it is crucial to utilize the *Apriori property*, which states that all non-empty subsets of a partial periodic pattern must also be partial periodic patterns. This property allows for effective search space pruning, enabling efficient discovery of partial periodic patterns.

The Python code provided below demonstrates how to implement the discovery of partial periodic patterns using the PAMI package. The `PPPGrowth` algorithm is employed to identify patterns that meet the specified periodic support criteria.

Program 7: Finding Partial Periodic Patterns

```python
from PAMI.partialPeriodicPattern.basic import PPPGrowth as alg

# Initialize the PPPGrowth algorithm with necessary parameters
obj = alg.PPPGrowth(iFile='Temporal_T10I4D100K.csv', minPS=100,
↪  period=200, sep='\t')

# Mine the partial periodic patterns
obj.mine()

# Save the patterns to a file
obj.save('partialPeriodicPatterns.txt')

# Retrieve the patterns as a DataFrame
PPFPs = obj.getPatternsAsDataFrame()

# Display summary information
print('#Patterns: ' + str(len(PPFPs)))
print('Runtime: ' + str(obj.getRuntime()))
print('Memory (RSS): ' + str(obj.getMemoryRSS()))
print('Memory (USS): ' + str(obj.getMemoryUSS()))
```

6.6.3 Recurring Patterns

Recurring patterns [10] are a distinct subset of periodic patterns that display periodic behavior within specific time intervals. These patterns are beneficial when certain items or itemsets are consistently purchased during particular periods, such as specific hours of the day, days of the week, or seasons of the year.

For example, consider a recurring pattern like $\{greenTea, Obento\}$ $\{[11:00, 14:00], [16:00, 21:00]\}$. This pattern indicates that the combination of green tea and Obento is frequently purchased during lunch (11:00 AM to 2:00 PM) and dinner (4:00 PM to 9:00 PM) time intervals. Such patterns are valuable for businesses to understand customer behavior and optimize inventory or promotional strategies during peak hours.

The following Python code demonstrates using the PAMI package to identify recurring patterns within a temporal database.

Program 8: Finding Recurring Patterns

```python
from PAMI.recurringPattern.basic import RPGrowth as alg

# Initialize the RPGrowth algorithm with appropriate parameters
obj = alg.RPGrowth(iFile='Temporal_T10I4D100K.csv', minPS=20, maxPer=100, minRec=1, sep='\t')  # Separator used in the data file

# Mine the recurring patterns
obj.mine()

# Save the patterns to a file
obj.save('recurringPatterns.txt')

# Retrieve the patterns as a DataFrame
recurringPatterns = obj.getPatternsAsDataFrame()

# Display summary information
print('#Patterns: ' + str(len(recurringPatterns)))
print('Runtime: ' + str(obj.getRuntime()))
print('Memory (RSS): ' + str(obj.getMemoryRSS()))
print('Memory (USS): ' + str(obj.getMemoryUSS()))
```

6.7 Conclusion

Periodic-frequent pattern mining is crucial for uncovering trends in temporal data, but it faces challenges such as the rare item problem and the inability to identify partially periodic patterns. The rare item problem stems from the assumption of uniform item frequencies, which can either lead to missing important patterns or generating excessive, irrelevant results. To overcome this, periodic-correlated pattern mining introduces measures like all-confidence and periodic-all-confidence to balance the influence of frequent and rare items, ensuring that valuable patterns are not overlooked. Additionally, traditional models struggle to capture patterns with intermittent periodicity, which partial periodic pattern discovery addresses by relaxing strict periodicity requirements. These advancements enhance the ability to extract meaningful insights from complex temporal datasets, supporting more effective decision-making and strategy optimization. As the field evolves, ongoing innovations will continue to refine pattern mining techniques, driving progress in data analysis.

References

1. Syed Khairuzzaman Tanbeer, Chowdhury Farhan Ahmed, Byeong-Soo Jeong, and Young-Koo Lee. 2009. Discovering Periodic-Frequent Patterns in Transactional Databases. In Proceedings of the 13th Pacific-Asia Conference on Advances in Knowledge Discovery and Data Mining. Springer-Verlag, Berlin, Heidelberg, 242–253.
2. R. Uday Kiran, P. Krishna Reddy: Towards Efficient Mining of Periodic-Frequent Patterns in Transactional Databases. DEXA (2) 2010: 194–208
3. Kiran, R. U., Veena, P., Ravikumar, P., Saideep, C., Zettsu, K., Shang, H., Toyoda, M., Kitsuregawa, M., and Reddy, P. K. (2022). Efficient Discovery of Partial Periodic Patterns in Large Temporal Databases. Electronics, 11(10), 1523.
4. Pamalla Veena, Rage Uday Kiran, Penugonda Ravikumar, Likhitha Palla, Yuto Hayamizu, Kazuo Goda, Masashi Toyoda, Koji Zettsu, Sourabh Shrivastava: A fundamental approach to discover closed periodic-frequent patterns in very large temporal databases. Appl. Intell. 53(22): 27344–27373 (2023).
5. Palla Likhitha, Pamalla Veena, R. Uday Kiran, Yutaka Watanobe, Koji Zettsu: Discovering Maximal Partial Periodic Patterns in Very Large Temporal Databases. IEEE BigData 2021: 1460–1469.
6. Palla Likhitha, Penugonda Ravikumar, Deepika Saxena, Rage Uday Kiran, Yutaka Watanobe: k-PFPMiner: Top-k Periodic Frequent Patterns in Big Temporal Databases. IEEE Access 11: 119033–119044 (2023).
7. J. N. Venkatesh, R. Uday Kiran, P. Krishna Reddy, Masaru Kitsuregawa: Discovering Periodic-Correlated Patterns in Temporal Databases. Trans. Large Scale Data Knowl. Centered Syst. 38: 146–172 (2018).
8. R. Uday Kiran, P. Krishna Reddy: An Alternative Interestingness Measure for Mining Periodic-Frequent Patterns. DASFAA (1) 2011: 183–192.
9. Pamalla Veena, Rage Uday Kiran, Penugonda Ravikumar, Likhitha Palla, Yutaka Watanobe, Sadanori Ito, Koji Zettsu, Masashi Toyoda, B. V. V. Raj: 3P-ECLAT: mining partial periodic patterns in columnar temporal databases. Appl. Intell. 54(11–12): 657–679 (2024)
10. R. Uday Kiran, Haichuan Shang, Masashi Toyoda, Masaru Kitsuregawa: Discovering Recurring Patterns in Time Series. EDBT 2015: 97–108.

Chapter 7
Spatial Databases: Representation, Creation, and Statistics

Abstract This chapter provides a comprehensive guide to working with geo-referenced databases, focusing on both transactional and temporal formats. It covers the theoretical foundations, including formal definitions and mathematical representations, as well as practical applications, such as generating synthetic datasets, converting dataframes, and analyzing statistical details. Using the PAMI package, users can create large-scale geo-referenced databases tailored to specific requirements, convert existing data into spatially and temporally aware formats, and derive key statistical insights. This chapter equips data scientists and researchers with the tools and knowledge to effectively manage and analyze spatial data.

7.1 Introduction

A spatial database, also known as a structured certain binary spatial database, stores data with spatial attributes, such as the position of pixels in raster images or the locations of points, lines, and polygons in vector images. Figure 7.1 visually represents the complex factors involved in forming a spatial database. This figure highlights the intricate relationships and interactions crucial for organizing and analyzing temporal data within the spatial database framework.

A spatial database does not function independently; it requires integration with other types of databases to leverage its capabilities thoroughly. Typically, the data about spatial items (or objects) is modeled as part of a transactional or temporal database. When a spatial database is combined with a transactional database, the resultant is called a *geo-referenced transactional database*. Similarly, when a spatial database is integrated with a temporal database, it forms a *geo-referenced temporal database*. This chapter provides an in-depth exploration of spatial databases, as well as geo-referenced transactional and geo-referenced temporal databases, offering insights into their structures, functionalities, and applications.

This chapter covers the following key aspects of transactional databases:

1. **Theoretical Representation**: It provides a formal definition of spatial databases, geo-referenced transactional databases, and geo-referenced temporal databases

U. K. Rage, *Hands-on Pattern Mining*,
https://doi.org/10.1007/978-981-96-6791-8_7

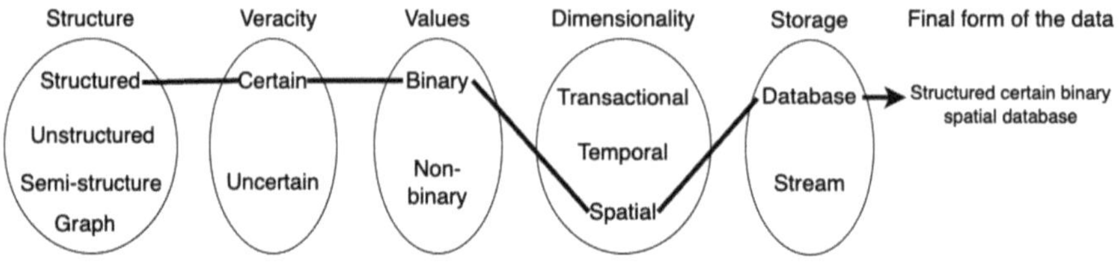

Fig. 7.1 Factors contributing to the creation of a spatial database

using set theory, laying the theoretical foundation for understanding these complex systems.
2. **Practical Representation**: The chapter explores how these databases are practically implemented and stored in computer systems, offering insights into their real-world applications and management.
3. **Synthetic Database Creation**: It discusses techniques for generating synthetic databases, which are crucial for testing, benchmarking, and evaluating the performance of various pattern mining algorithms.
4. **Dataframe Conversion**: The chapter outlines methods for converting structured dataframes into geo-referenced transactional and geo-referenced temporal databases, enhancing their utility for broader data analysis and application development.
5. **Database Statistics**: It explains how to derive and interpret statistical details about the databases, providing tools for assessing their characteristics and performance.

7.2 Theoretical Representation

7.2.1 Spatial Database

A spatial database represents a collection of items along with their respective coordinates. Each item must have a unique name, and no two items can have the same coordinates. The formal definition is provided below.

Definition 7.1 (Spatial Database) Let $SI = \{i_1, i_2, \cdots, i_n\}$, $n \geq 1$, be a set of spatial items. Let $P_{i_j} = \{(x_1, y_1), (x_2, y_2), \cdots, (x_p, y_p)\}$, $p \geq 1$, denote the set of coordinates for an item $i_j \in SI$. The location (or spatial) database SD is a set of items and their coordinates. That is, $SD = \{(i_1, P_{i_1}), (i_2, P_{i_2}), \cdots, (i_n, P_{i_n})\}$. This definition allows the spatial database to represent items of various spatial forms, such as pixels, points, lines, or polygons.

Example 7.1 Let $SI = \{a, b, c, d, e, f, g\}$ be a set of spatial items, each representing a sensor at specific coordinates. Figure 7.2a displays the spatial database for all items in SI. The spatial visualization of these items within a coordinate system is shown in Fig. 7.2b.

Fig. 7.2 A spatial database. (**a**) Example of spatial items and their coordinates. (**b**) Visualization of the spatial locations of the items

Item	Coordinates
a	*POINT(0 1)*
b	*POINT(2 1)*
c	*POINT(1 0)*
d	*POINT(1 2)*
e	*POINT(1 1)*
f	*POINT(1 5)*
g	*POINT(1 6)*

(a) Spatial database

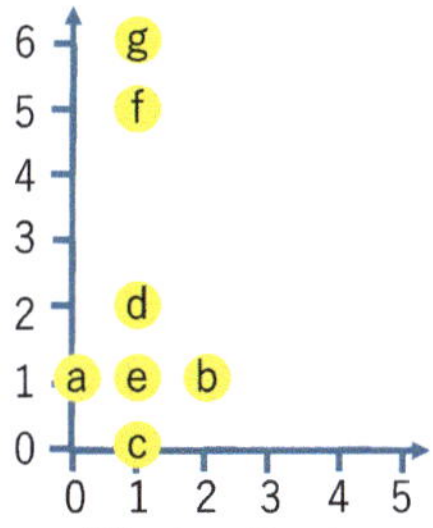

(b) visualization

7.2.2 *Geo-referenced Transactional Database*

When the data of stationary spatial objects is stored in a transactional database format, the resulting system is referred to as a geo-referenced transactional database [1]. This type of database integrates the spatial characteristics of objects with the transactional data, enabling the analysis and management of spatial information within the context of transactional processes.

Definition 7.2 (Geo-referenced Transactional Database) Let $X \subseteq SI$ be an itemset (or a pattern). If X contains k items, where $k \geq 1$, then X is called a k-pattern. A transaction $t_{tid} = (tid, Y)$, where $tid \geq 1$ represents the transaction identifier, and $Y \subseteq SI$ is a pattern. A **transactional database**, denoted as TDB, is a collection of transactions, defined as $TDB = \{t_1, t_2, \cdots, t_m\}$, where $1 \leq m \leq |TDB|$, and $|TDB|$ represents the size of the database. The combination of a spatial database and a transactional database forms a geo-referenced transactional database, denoted as GTD. Formally, $GTD = SD \times TDB$.

Example 7.2 Figure 7.3a illustrates a spatial database, while Fig. 7.3b presents a hypothetical transactional database generated by the spatial items. This database contains seven transactions, identified by transaction identifiers (or tid) numbered 1 to 7. When the item information in the transactional database is replaced with their corresponding spatial information, the resultant database is termed a geo-referenced transactional database. The horizontal format of this database is shown in Fig. 7.3c, and the vertical format is depicted in Fig. 7.4.

> **Important**

A geo-referenced transactional database contains stationary spatial items whose positions do not vary over time.

Item	Coordinates
a	POINT(0 1)
b	POINT(2 1)
c	POINT(1 0)
d	POINT(1 2)
e	POINT(1 1)
f	POINT(1 5)
g	POINT(1 6)

(a)

×

tid	Items
1	*abc*
2	*acd*
3	*bdef*
4	*acef*
5	*aefg*
6	*acef*
7	*acef*

(b)

=

tid	Items
1	POINT(0 1) POINT(2 1) POINT(1 0)
2	POINT(0 1) POINT(1 0) POINT(1 2)
3	POINT(2 1) POINT(1 2) POINT(1 1) POINT (1 5)
4	POINT(0 1) POINT(1 0) POINT(1 1) POINT (1 5)
5	POINT(0 1) POINT(1 1) POINT (1 5) POINT(1 6)
6	POINT(0 1) POINT(1 0) POINT(1 1) POINT(1 5)
7	POINT(0 1) POINT(1 0) POINT(1 1) POINT(1 5)

(c)

Fig. 7.3 Creation of a geo-referenced transactional database: (**a**) spatial database, (**b**) transactional database, and (**c**) geo-referenced transactional database

tid	POINT(0 1)	POINT(2 1)	POINT(1 0)	POINT(1 2)	POINT(1 1)	POINT(1 5)	POINT(1 6)
1	1	1	1	0	0	0	0
2	1	0	1	1	0	0	0
3	0	1	0	1	1	1	0
4	1	0	1	0	1	1	0
5	1	0	0	0	1	1	1
6	1	0	1	0	1	1	0
7	1	0	1	0	1	1	0

Fig. 7.4 Geo-referenced transactional database in the vertical format

7.2.3 Geo-referenced Temporal Database

If the data of the spatial objects is stored in a temporal database format, the resulting database is known as a *geo-referenced temporal database*.

Definition 7.3 (Geo-referenced Temporal Database) A transaction, denoted as t_{tid}, is a triplet containing a transaction identifier, a timestamp, and a pattern. That is, $t_{tid} = (tid, ts, Y)$, where $tid \geq 1$ represents the transaction identifier, $ts \in \mathbb{R}^+$ represents the timestamp, and $Y \subseteq SI$ is a pattern. A **temporal database**, denoted as $TempDB$, is an ordered collection of transactions by time. That is, $TempDB = \{t_1, t_2, \cdots, t_m\}$, where $1 \leq m \leq |TempDB|$, and $|TempDB|$ represents the size of the database. Integrating spatial and temporal databases results in a geo-referenced temporal database denoted as $GTempD$. Formally, $GTempD = SD \times TempDB$.

Example 7.3 Figure 7.5a shows the spatial database. Figure 7.5b depicts a hypothetical temporal database generated by the spatial items. This database includes seven transactions, each numbered from 1 to 7 as the transaction identifiers (or tid). The database is characterized by irregular temporal intervals, indicating nonuniform gaps between consecutive transactions. When the item information in the temporal database is replaced with their spatial information, the resulting database is known as a geo-referenced temporal database. The horizontal format of this database is shown in Fig. 7.5c, and the vertical format is illustrated in Fig. 7.6.

Item	Coordinates
a	POINT(0 1)
b	POINT(2 1)
c	POINT(1 0)
d	POINT(1 2)
e	POINT(1 1)
f	POINT(1 5)
g	POINT(1 6)

(a)

tid	Items
1	*a b c*
2	*a c d*
3	*b d e f*
4	*a c e f*
5	*a e f g*
6	*a c e f*
7	*a c e f*

(b)

tid	Items
1	POINT(0 1) POINT(2 1) POINT(1 0)
2	POINT(0 1) POINT(1 0) POINT(1 2)
3	POINT(2 1) POINT(1 2) POINT(1 1) POINT (1 5)
4	POINT(0 1) POINT(1 0) POINT(1 1) POINT (1 5)
5	POINT(0 1) POINT(1 1) POINT (1 5) POINT(1 6)
6	POINT(0 1) POINT(1 0) POINT(1 1) POINT(1 5)
7	POINT(0 1) POINT(1 0) POINT(1 1) POINT(1 5)

(c)

Fig. 7.5 Creation of a geo-referenced temporal database: (**a**) spatial database, (**b**) temporal database, and (**c**) geo-referenced temporal database

tid	ts	POINT(0 1)	POINT(2 1)	POINT(1 0)	POINT(1 2)	POINT(1 1)	POINT(1 5)	POINT(1 6)
1	1	1	1	1	0	0	0	0
2	2	1	0	1	1	0	0	0
3	3	0	1	0	1	1	1	0
4	3	1	0	1	0	1	1	0
5	6	1	0	0	0	1	1	1
6	6	1	0	1	0	1	1	0
7	8	1	0	1	0	1	1	0

Fig. 7.6 The vertical format of a geo-referenced temporal database

Important

A geo-referenced temporal database is a temporal database containing spatial items.

7.3 Practical Representation

7.3.1 Spatial Database

To create a spatial database, follow these rules:

1. **One Transaction per Line**: Each line in the file should represent a unique transaction. No two lines should be identical.
2. **Two Columns per Line**: Each line must have exactly two columns. A delimiter should separate these columns. By default, the PAMI algorithms use a `tab` as the delimiter, but you can also use commas or spaces.
3. **Order of Elements in a Line**: The first column should contain the name of the item. The second column should include the spatial information corresponding to that item.

4. **No Duplicates**: Each column must have unique entries. In other words, no two rows in a column should have the same value.

In summary, the format of a spatial database should be "$spatialItem\langle sep\rangle coordinates$." For example, if using a tab as the delimiter, the spatial database shown in Fig. 7.2a would look like this:

```
a	POINT(0 1)
b	POINT(2 1)
c	POINT(1 0)
d	POINT(1 2)
e	POINT(1 1)
f	POINT(1 5)
g	POINT(1 6)
```

7.3.2 Geo-referenced Transactional Database

A geo-referenced transactional database is essentially a transactional database that contains spatial items. It follows all the rules for a transactional database (see Sect. 3.3). In addition to these rules, items should be replaced with their spatial coordinates. The format for this type of database is

$$coordinates_1\langle sep\rangle coordinates_2\langle sep\rangle coordinates_3\langle sep\rangle \cdots$$

If using a `tab` as the delimiter, the geo-referenced transactional database shown in Fig. 7.3c would appear like this:

```
POINT(0 1)	POINT(2 1)	POINT(1 0)
POINT(0 1)	POINT(1 0)	POINT(1 2)
POINT(2 1)	POINT(1 2)	POINT(1 1)	POINT (1 5)
POINT(0 1)	POINT(1 0)	POINT(1 1)	POINT (1 5)
POINT(0 1)	POINT(1 1)	POINT(1 5)	POINT(1 6)
POINT(0 1)	POINT(1 0)	POINT(1 1)	POINT(1 5)
POINT(0 1)	POINT(1 0)	POINT(1 1)	POINT(1 5)
```

7.3.3 Geo-referenced Temporal Database

A geo-referenced temporal database [2] is a temporal database that includes spatial items. It adheres to all the rules for a temporal database (see Sect. 5.3). Additionally, items should be replaced with their spatial coordinates. The format for this type of database is

$$timestamp\langle sep\rangle coordinates_1\langle sep\rangle coordinates_2\langle sep\rangle coordinates_3\langle sep\rangle \cdots$$

If the delimiter is a tab, the geo-referenced temporal database shown in Fig. 7.5c would look like this:

```
1   POINT(0 1)  POINT(2 1)  POINT(1 0)
2   POINT(0 1)  POINT(1 0)  POINT(1 2)
3   POINT(2 1)  POINT(1 2)  POINT(1 1)  POINT (1 5)
3   POINT(0 1)  POINT(1 0)  POINT(1 1)  POINT (1 5)
6   POINT(0 1)  POINT(1 1)  POINT(1 5)  POINT(1 6)
6   POINT(0 1)  POINT(1 0)  POINT(1 1)  POINT(1 5)
8   POINT(0 1)  POINT(1 0)  POINT(1 1)  POINT(1 5)
```

7.4 Creating Synthetic Datasets

The PAMI package provides a robust and versatile tool for generating synthetic geo-referenced transactional and temporal databases to meet various requirements. Each item in these databases is assigned a unique random spatial coordinate within a defined range. This range is specified by the intervals (x_1, y_1) and (x_2, y_2), where $x_1 \leq x_2$ and $y_1 \leq y_2$. Figure 7.7 illustrates the area within which these random coordinates will be assigned to the items in the database.

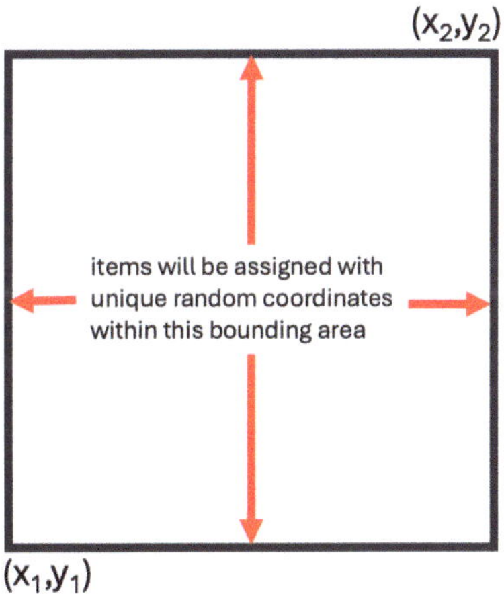

Fig. 7.7 The area within which random coordinates will be assigned to the items in the database

7.4.1 Generating Synthetic Geo-referenced Transactional Database

The PAMI package offers algorithms to generate synthetic geo-referenced transactional databases based on user specifications. Users can create a database of any size, with items having coordinates within specified intervals.

To illustrate, consider the following sample code that generates a synthetic geo-referenced transactional database with 100,000 transactions. Each transaction includes an average of 10 items selected from a set of 1000 possible items, with coordinates ranging from (0,0) to (100,100):

Program 1: Generating Synthetic Geo-referenced Transactional Database

```
from PAMI.extras.syntheticDataGenerator import
↪   GeoReferentialTransactionalDatabase as db

obj =
↪   db.GeoReferentialTransactionalDatabase(databaseSize=100000,
↪   avgItemsPerTransaction=10, numItems=1000, x1=0, y1=0,
↪   x2=100, y2=100, sep='\t')
obj.create()
obj.save('geoReferencedTransactionalDatabase.csv')
#read the generated transactions into a dataframe
GRTDF=obj.getTransactions()
print('Runtime: ' + str(obj.getRuntime()))
print('Memory (RSS): ' + str(obj.getMemoryRSS()))
print('Memory (USS): ' + str(obj.getMemoryUSS()))
```

7.4.2 Generating Synthetic Geo-referenced Temporal Database

The PAMI package provides algorithms to create synthetic geo-referenced temporal databases tailored to user specifications. Users can generate databases of any size, with transactions occurring regularly or irregularly and items having coordinates within a defined interval.

You can create a synthetic geo-referenced temporal database using the following sample code. This example generates a database with 100,000 transactions, each containing an average of 10 items from a set of 1000 possible items. The coordinates for the items are within the range (0,0) to (100,100). The code also specifies probabilities for timestamps to illustrate their distribution:

Program 2: Generating Synthetic Geo-referenced Temporal Database

```
from PAMI.extras.syntheticDataGenerator import
    GeoReferentialTemporalDatabase as db

obj = db.GeoReferentialTemporalDatabase(databaseSize=100000,
    avgItemsPerTransaction=10, numItems=1000,
    occurrenceProbabilityOfSameTimestamp=0,
    occurrenceProbabilityToSkipSubsequentTimestamp=0, x1=0,
    y1=0, x2=100, y2=100, sep='\t')
obj.create()
obj.save('geoReferentialTemporalDatabase.csv')
GRTempDF=obj.getTransactions()
print('Runtime: ' + str(obj.getRuntime()))
print('Memory (RSS): ' + str(obj.getMemoryRSS()))
print('Memory (USS): ' + str(obj.getMemoryUSS()))
```

7.5 Deriving Geo-referenced Databases from a Dataframe

The PAMI package provides functionality to convert a dataframe into either a geo-referenced transactional database or a geo-referenced temporal database, making it suitable for transaction-based data analysis.

7.5.1 Dataframe to Geo-referenced Transactional Database

The following code demonstrates how to convert a dataframe into a geo-referenced transactional database:

Program 3: Converting a Dataframe into a Geo-referenced Transactional Database

```
from PAMI.extras.convert import DF2DB as alg
import pandas as pd
import numpy as np

#creating a 4 x 4 dataframe with random values
data = np.random.randint(1, 100, size=(4, 4))
```

```
dataFrame = pd.DataFrame(data,
columns=['POINT(0 0)', 'POINT(0 1)', 'POINT(0 2)', 'POINT(0
↪  3)'])

obj = alg.DF2DB(dataFrame)
obj.convert2TransactionalDatabase(
oFile='georeferencedTransactionalDatabase.txt',
condition='>=', thresholdValue=36
)
print('Runtime: ' + str(obj.getRuntime()))
print('Memory (RSS): ' + str(obj.getMemoryRSS()))
print('Memory (USS): ' + str(obj.getMemoryUSS()))
```

7.5.2 Dataframe to Geo-referenced Temporal Database

The following code demonstrates how to convert a dataframe into a geo-referenced temporal database:

Program 4: Converting a Dataframe into a Geo-referenced Temporal Database

```
from PAMI.extras.convert import DF2DB as alg
import pandas as pd
import numpy as np

#creating a 5 x 4 dataframe with random values
data = np.random.randint(1, 100, size=(5, 4))
dataFrame = pd.DataFrame(data,
                columns=['POINT(0 0)', 'POINT(0 1)',
                'POINT(0 2)', 'POINT(0 3)'])
# Adding a timestamp column with specific values
timestamps = [1, 3, 3, 5, 8]
dataFrame.insert(0, 'timestamp', timestamps)
#converting the database into a georeferenced temporal database
obj = alg.DF2DB(dataFrame)
obj.convert2TemporalDatabase(
          oFile='georeferencedTemporalDatabase.txt',
         condition='>=', thresholdValue=36)
print('Runtime: ' + str(obj.getRuntime()))
```

```
print('Memory (RSS): ' + str(obj.getMemoryRSS()))
print('Memory (USS): ' + str(obj.getMemoryUSS()))
```

7.6 Knowing the Statistical Details

The `dbStats` sub-sub-package in the `extras` sub-package of PAMI provides users with statistical details about a geo-referenced database.

7.6.1 Statistical Details of a Geo-referenced Transactional Database

The PAMI library provides the following statistical details for a geo-referenced transactional database:

1. **Database size**: Total number of transactions
2. **Total number of items**: Unique items in the database
3. **Transaction lengths**: Minimum, average, and maximum number of items per transaction
4. **Standard deviation of transaction sizes**: Variability in the number of items per transaction
5. **Variance in transaction sizes**: Measure of dispersion in transaction sizes
6. **Sparsity**: Measure of how sparse the data is
7. **Item frequencies**: Count of each item's occurrence in the database
8. **Distribution of transaction lengths**: How transaction sizes are distributed across the database
9. **Spatial visualization**: Visual representation of item locations

Here is how to use `dbStats` to obtain these statistics:

Program 5: Deriving Statistical Details for a Geo-referenced Transactional Database

```
from PAMI.extras.dbStats import
↪   GeoreferencedTransactionalDatabase as stat

obj = stat.GeoreferencedTransactionalDatabase(iFile =
↪   "georeferencedTransactionalDatabase.txt")
obj.run()
```

```
obj.printStats()
obj.plotGraphs()
```

7.6.2 Statistical Details of a Geo-referenced Temporal Database

The PAMI library provides the following statistical details for a geo-referenced temporal database:

1. **Database size**: The total number of transactions in the database
2. **Total number of items**: The number of unique items in the database
3. **Transaction lengths**: The minimum, average, and maximum number of items in the transactions
4. **Standard deviation of transaction sizes**: A measure of the variability in the number of items per transaction
5. **Variance in transaction sizes**: A measure of how transaction sizes differ from the average
6. **Sparsity**: The proportion of empty (zero) elements in the database
7. **Item frequencies**: The count of occurrences of each item in the database
8. **Distribution of transaction lengths**: How transaction sizes are spread across the database
9. **Inter-arrival times**: The minimum, average, and maximum time intervals between transactions
10. **Periodicity of items**: The minimum, average, and maximum time intervals between occurrences of the same item
11. **Spatial visualization**: Visual representation of item locations

Here is an example of how to use the `dbStats` to obtain the statistics:

Program 6: Deriving the Statistical Details for Geo-referenced Temporal Database

```
from PAMI.extras.dbStats import GeoreferencedTemporalDatabase as stat

obj = stat.GeoreferencedTemporalDatabase(iFile = "georeferencedTemporalDatabase.txt")
obj.run()
obj.printStats()
obj.plotGraphs()
```

7.7 Conclusion

In this chapter, we explored geo-referenced databases' creation, manipulation, and analysis, focusing on both transactional and temporal contexts. We delved into the practical steps for generating synthetic datasets, providing a hands-on approach to creating large-scale geo-referenced transactional and temporal databases using the PAMI package. Additionally, we examined methods to convert existing dataframes into geo-referenced formats, thus enhancing their applicability in spatial-temporal data analysis.

Finally, we discussed the statistical analysis of these databases, highlighting the importance of understanding key metrics such as transaction lengths, sparsity, and periodicity. By leveraging the methods provided in the PAMI package, users can efficiently derive and visualize these statistics, enabling more informed decision-making in spatial-temporal data management and analysis.

This chapter serves as a comprehensive guide for anyone working with geo-referenced databases. It provides both theoretical foundations and practical applications to empower data scientists and researchers.

References

1. R. Uday Kiran, Sourabh Shrivastava, Philippe Fournier-Viger, Koji Zettsu, Masashi Toyoda, Masaru Kitsuregawa: Discovering Frequent Spatial Patterns in Very Large Spatiotemporal Databases. SIGSPATIAL/GIS 2020: 445–448.
2. Palla Likhitha, Pamalla Veena, Rage Uday Kiran, Koji Zettsu: Discovering Geo-referenced Frequent Patterns in Uncertain Geo-referenced Transactional Databases. PAKDD (3) 2023: 29–41.

Chapter 8
Pattern Discovery in Spatial Databases

Abstract This chapter presents a comprehensive approach to mining geo-referenced frequent patterns and geo-referenced periodic-frequent patterns by integrating spatial and temporal dimensions in transactional databases. Geo-referenced frequent patterns focus on identifying sets of spatially proximate items that occur frequently, while geo-referenced periodic-frequent patterns extend this by considering periodicity in their occurrence. Efficient search techniques such as the anti-monotonic property and neighborhood-aware depth-first search are utilized to manage the large search space inherent in these tasks. The chapter also introduces algorithms from the PAMI library, including FSP-growth and GPFPMiner, which facilitate the discovery of these patterns. Real-world applications, such as environmental sensor networks, can benefit from the insights gained through these mining techniques, enabling a better understanding of spatial-temporal dynamics. Practical Python implementations are provided to demonstrate how to mine, save, and analyze geo-referenced patterns in large datasets. This chapter highlights the importance of combining spatial and temporal analyses for improving decision-making and system optimization in various domains.

8.1 Introduction

The previous chapter explored the concepts of spatial database construction, representation, and statistical analysis. Building on that foundation, this chapter focuses on extracting and analyzing meaningful patterns, which are critical for understanding trends and behaviors over space and time.

Traditional frequent pattern mining and its variants, such as correlated pattern mining and periodic-frequent pattern mining, generally assume that the spatial relationships between items do not affect the overall interestingness of a pattern. However, this assumption limits the effectiveness of these models when applied to spatial databases. In many real-world applications, patterns whose items are spatially close to one another are often more significant to users than those where the items are widely dispersed across a coordinate system. Consequently, incorporating spatial proximity into the pattern mining process is essential for uncovering more meaningful insights from spatiotemporal datasets.

U. K. Rage, *Hands-on Pattern Mining*,
https://doi.org/10.1007/978-981-96-6791-8_8

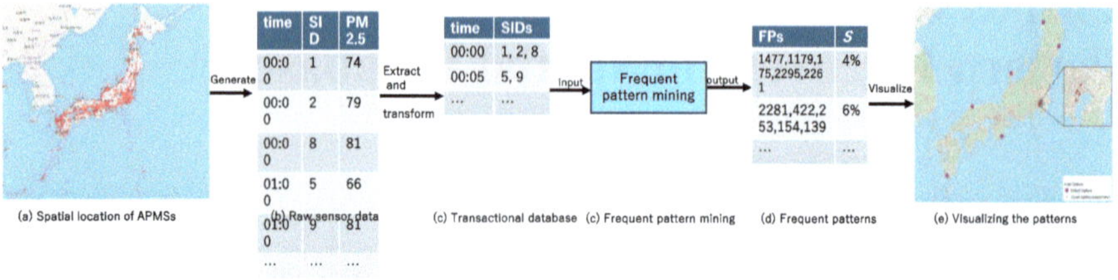

Fig. 8.1 Air pollution analytics using frequent patterns. The terms "SID," "PM2.5," "FPs," and "*S*" represent "station identifier," "particular matter with diameter 2.5 μm or less," "frequent patterns," and "*support*," respectively

Example 8.1 Air pollution is a significant contributor to many cardio-respiratory health issues reported by residents of Japan. To address this, the Japanese Ministry of the Environment established the Atmospheric Environmental Regional Observation System (AEROS) (https://soramame.env.go.jp/). This system consists of air pollution sensors distributed across the country. Figure 8.1a shows the spatial distribution of these sensors in AEROS. The data generated by this sensor network at hourly intervals (see Fig. 8.1b) can be structured as a transactional database (Fig. 8.1c) and analyzed using the frequent pattern model (Fig. 8.1d) to identify sets of sensors (or geographic regions) where high levels of an air pollutant, such as $PM_{2.5}$,[1] occur frequently.

For instance, let $\{365, 996, 2769, 4815, 5687, 2395\}$[2] and $\{4276, 4341, 4495, 4273, 4455\}$[3] be two frequent patterns identified in the air pollution database. The frequent pattern model treats both equally relevant, regardless of the spatial distances between the sensors. However, the user might find the latter pattern more meaningful, as it corresponds to a specific geographical area (i.e., the bay areas south of Tokyo) where residents have been consistently exposed to high levels of $PM_{2.5}$. This highlights a limitation of applying traditional frequent pattern models on geo-referenced transactional databases, where spatial information is an integral part of the data.

To discover valuable patterns in spatial (or geo-referenced) datasets, the researchers exploited the notion of "neighborhood items" and introduced different types of interesting patterns hidden in geo-referenced transactional and temporal databases. This chapter describes the notion of "neighborhood items," how to create a neighborhood file, and discuss various types of patterns that can be discovered from geo-referenced databases.

---

[1] $PM_{2.5}$ refers to fine particulate matter with a diameter of 2.5 μm or smaller.

[2] This pattern is represented by black dots in Fig. 8.1e.

[3] This pattern is represented by red dots in Fig. 8.1e.

This chapter delves into the following topics:

1. **Neighboring Items**: We will define the notion of neighboring items, how to create a neighborhood file, and describe its process.
2. **Pattern Discovery in Geo-referenced Transactional Databases**: This subsection describes the model of geo-referenced frequent patterns and how to find them using the PAMI package.
3. **Pattern Discovery in Geo-referenced Temporal Databases**: This subsection describes the model of geo-referenced periodic patterns and how to find them using the PAMI package.

8.2 Neighboring Items

The concept of "neighborhood" is a crucial differentiator between pattern discovery in spatial databases and conventional pattern discovery in transactional or temporal databases. In spatial databases, such as geo-referenced transactional and geo-referenced temporal databases, the goal is to uncover patterns consisting only of neighboring items. Neighboring items are those located close to one another in space. Next, we will formally define the notion of "neighboring items."

8.2.1 Definition

Definition 8.1 (Neighborhood Items) Two spatial items, i_p and $i_q \in SI$, are considered **neighbors** if the distance between them, $Dist(i_p, i_q) = Dist(i_q, i_p)$, is less than or equal to a user-defined *maximum distance* ($maxDist$). Here, $Dist(.)$ is a distance function that adheres to the commutative property. The set of all neighboring items for a given item $i_j \in I$ is denoted by N_{i_j}.

Example 8.2 Let $I = \{POINT(0, 1), POINT(2, 1), POINT(1, 0), POINT(1, 2),$- $POINT(1, 1), POINT(1, 5), POINT(1, 6)\}$ be the set of locations of spatial items. The spatial database for all items in I is presented in Table 8.1. Using the Euclidean distance as the distance function, the distance between items $POINT(0, 1)$ and $POINT(1, 0)$ is given by $Dist(POINT(0, 1), POINT(1, 0)) = 1.414$, since the user-specified $maxDist = 1.5$, $POINT(0, 1)$, and $POINT(1, 0)$ are considered neighbors because $Dist(POINT(0, 1), POINT(1, 0)) \leq maxDist$. Additionally, items $POINT(1, 2)$ and $POINT(1, 1)$ are also neighbors of $POINT(0, 1)$, resulting in the set of neighbors $N_{POINT(0,1)} = \{POINT(1, 0), POINT(1, 2), POINT(1, 1)\}$. The complete list of neighbors for every item in the database is shown in Table 8.2.

Table 8.1 Spatial database

Items	Items	Items	Items
$POINT(0, 1)$	$POINT(1, 0)$	$POINT(1, 1)$	$POINT(1, 6)$
$POINT(2, 1)$	$POINT(1, 2)$	$POINT(1, 5)$	

Table 8.2 Neighborhood items

Item	Neighbors
$POINT(0, 1)$	$\{POINT(1, 0), POINT(1, 2), POINT(1, 1)\}$
$POINT(2, 1)$	$\{POINT(1, 0), POINT(1, 2), POINT(1, 1)\}$
$POINT(1, 0)$	$\{POINT(0, 1), POINT(2, 1), POINT(1, 1)\}$
$POINT(1, 2)$	$\{POINT(0, 1), POINT(2, 1), POINT(1, 1)\}$
$POINT(1, 1)$	$\{POINT(0, 1), POINT(2, 1), POINT(1, 0), POINT(1, 2)\}$
$POINT(1, 5)$	$\{POINT(1, 6)\}$
$POINT(1, 6)$	$\{POINT(1, 5)\}$

8.2.2 Practical Representation

To create a neighborhood file for the items, follow these rules:

1. **One Transaction per Line**: Each line in the file should represent a unique transaction. No two lines should be identical.
2. **Order of Elements in a Line**: The first element in a line represents the main item. The remaining elements in a line represent the neighbors of the main item.
3. **Delimiter**: A delimiter should separate the items in a line. By default, the PAMI algorithms use a `tab` as the delimiter, but you can also use commas or spaces.

In summary, the format of a neighborhood file should be

$$spatialItem_1\langle sep\rangle spatialItem_2\langle sep\rangle spatialItem_3\langle sep\rangle \cdots$$

For example, if using a `tab` as the delimiter, the neighborhood file shown in Table 8.2(a) would look like this:

```
POINT(0,1)  POINT(1,0)  POINT(1,2)  POINT(1,1)
POINT(2,1)  POINT(1,0)  POINT(1,2)  POINT(1,1)
POINT(1,0)  POINT(0,1)  POINT(2,1)  POINT(1,1)
POINT(1,2)  POINT(0,1)  POINT(2,1)  POINT(1,1)
POINT(1,1)  POINT(0,1)  POINT(2,1)  POINT(1,0)  POINT(1,2)
POINT(1,5)  POINT(1,6)
POINT(1,6)  POINT(1,5)
```

8.2.3 Creating Neighborhood File

The PAMI package offers a utility for generating neighborhood files from geo-referenced transactional databases, where spatial items are represented as points. By leveraging Euclidean distance, users can efficiently identify neighbors for each spatial item within a specified maximum distance threshold.

The following Python code demonstrates how to process a geo-referenced transactional database. It finds the neighbors for each spatial item, restricted by the user-defined maximum distance, and outputs the results to a file or a dataframe for further analysis.

Program 1: Generating Neighborhood File

```
from PAMI.extras.neighbours import FindNeighboursUsingEuclidean
↪   as db

obj = db.FindNeighboursUsingEuclidean(
    iFile='spatiotransactional_T10I4D100K.csv',
    maxDist=10,
    sep='\t')
obj.create()
obj.save(oFile='neighbors.txt',)
#read the generated transactions into a dataframe
neighboringItems=obj.getNeighboringInformation()
#stats
print('Runtime: ' + str(obj.getRuntime()))
print('Memory (RSS): ' + str(obj.getMemoryRSS()))
print('Memory (USS): ' + str(obj.getMemoryUSS()))
```

This script identifies neighboring spatial items and provides runtime and memory statistics, facilitating performance evaluation and optimization for large datasets. The resulting dataframe can be seamlessly integrated into further analyses or visualizations.

8.3 Geo-referenced Frequent Pattern

8.3.1 The Basic Model

Let $J = \{j_1, j_2, \ldots, j_m\}$, where $m \geq 1$, represent a set of geo-referenced (or spatial) items. For each item $j_k \in J$, let $Q_{j_k} = \{(x_1, y_1), (x_2, y_2), \ldots, (x_q, y_q)\}$, $q \geq 1$, denote the set of coordinates associated with that item. The spatial database SD compiles these items along with their respective coordinates, such that:

$$SD = \{(j_1, Q_{j_1}), (j_2, Q_{j_2}), \ldots, (j_m, Q_{j_m})\}.$$

This structure allows for the representation of spatial items in various forms, including points, lines, or polygons. If $Y \subseteq J$ is an itemset (or pattern) containing r items, it is termed an r-pattern. A pattern Y in SD is defined as an interesting **geo-referenced pattern** if the maximum distance between any two of its items does not exceed the user-specified $maxDist$. Formally, Y is a geo-referenced pattern if:

$$\max(Dist(j_a, j_b) \mid \forall j_a, j_b \in Y) \leq maxDist, \quad \text{where } a, b \in [1, m],$$

and $Dist()$ is a distance function that satisfies the commutative property.

Example 8.3 Let $J = \{POINT(0, 1), POINT(2, 1), POINT(1, 0), POINT(1, 2),- POINT(1, 1), POINT(1, 5), POINT(1, 6)\}$ represent a set of air pollution measuring sensors (or their locations). Table 8.1 presents the spatial database for these items. Using the Euclidean distance function, the set of items $POINT(0, 1)$ and $POINT(1, 0)$, denoted as $\{POINT(0, 1), POINT(1, 0)\}$, forms a pattern containing two items, making it a 2-pattern. The distance between $POINT(0, 1)$ and $POINT(1, 0)$ is given by $Dist(POINT(0, 1), POINT(1, 0)) = 1.414$, since the user-specified $maxDist = 1.5$, $POINT(0, 1)$, and $POINT(1, 0)$ are neighbors, and thus $\{POINT(0, 1), POINT(1, 0)\}$ qualifies as an interesting geo-referenced pattern because $\max(Dist(x, z)) \leq maxDist$.

A transaction $t_{tid} = (tid, Y)$ consists of a transaction identifier $tid \geq 1$ and a pattern $Y \subseteq J$. A **transactional database**, denoted as TD, is a collection of such transactions: $TD = \{t_1, t_2, \ldots, t_n\}$, $1 \leq n \leq |TD|$, where $|TD|$ represents the size of the database. If a pattern $Z \subseteq Y$, it is said that Z occurs in transaction t_{tid}. Let $TID^Z = \{tid_a^Z, tid_b^Z, \ldots, tid_c^Z\}$, $a, b, c \in (1, |TD|)$ denote the set of all transaction identifiers where pattern Z appears in the database. The **support** of Z in TD, denoted as $sup(Z)$, represents the count of transactions containing Z, i.e., $sup(Z) = |TID^Z|$.

Definition 8.2 (Geo-referenced Frequent Pattern [1]) A pattern Z is considered a geo-referenced frequent pattern if it meets the following two conditions: (i) $sup(Z) \geq minSup$ and (ii) $\max(Dist(j_l, j_m) \mid \forall j_l, j_m \in Z) \leq maxDist$. Here, $minSup$ is the user-specified minimum support threshold.

Example 8.4 The transactional database for all items in Table 8.3 is shown in Table 8.4. **This model accommodates irregular transaction occurrences in a temporal database.** For instance, the first transaction indicates that sensors a, c, and d recorded hazardous levels of the air pollutant $PM_{2.5}$ at timestamp 1. Similar interpretations apply to the other transactions in Table 8.4. The size of this database is $n = |TD| = 14$. The spatial pattern ac appears in transactions with timestamps 1, 2, 3, 16, 17, 18, and 20, leading to $TS^{ac} = \{1, 2, 3, 16, 17, 18, 20\}$. Thus, the

Table 8.3 Spatial database

Item	Coord.
a	(0, 1)
b	(2, 1)
c	(1, 0)
d	(1, 2)
e	(1, 1)
f	(1, 5)
g	(1, 6)

Table 8.4 Transactional database

ts	Items
1	acd
2	$abce$
3	$abcd$
4	def
5	deg
8	adg
10	adf
11	bcd
12	adf
13	ae
16	$abcf$
17	$abcd$
18	$abcg$
20	$abcd$

support of ac in the database is $sup(ac) = |TS^{ac}| = 7$. Given the user-specified $minSup = 5$, the spatial pattern ac qualifies as a geo-referenced frequent pattern since $sup(ac) \geq minSup$.

Definition 8.3 (Problem Definition) Given a set of items J, a spatial database SD, a transactional database TD, a **minimum support** value $minSup$, and a **maximum distance** value $maxDist$, the **problem definition** is to identify all patterns in TD that have support no less than $minSup$ and a maximum distance between any two items no greater than $maxDist$.

8.3.2 *Handling the Search Space*

The search space of the geo-referenced frequent pattern is $2^n - 1$, where n represents the total number of items in the geo-referenced transactional database. One can handle this huge search space using the *anti-monotonic property* of $minSup$ and neighborhood-aware depth-first search. In the neighborhood-aware depth-first search, the depth-first search on the itemset lattice is carried only for the child nodes that contain all items as neighbors of the items in their respective parent nodes.

8.3.3 *Finding Geo-referenced Frequent Patterns*

The PAMI library provides FSP-growth and Spatial ECLAT algorithms to find geo-referenced frequent patterns. Below is an example Python script demonstrating how to find the geo-referenced frequent patterns using the FSP-growth algorithm.

Program 2: Finding Geo-referenced Frequent Patterns

```
from PAMI.georeferencedFrequentPattern.basic import FSPGrowth
↪  as alg

obj = alg.FSPGrowth("spatiotransactional_T10I4D100K.csv",
        "neighbors.txt", 1500, '\t')

obj.mine()
obj.save('georeferencedFrequentPatterns.txt')

# Retrieve the patterns as a DataFrame
GFPs = obj.getPatternsAsDataFrame()

# Display summary information
print('#Patterns: ' + str(len(GFPs)))
print('Runtime: ' + str(obj.getRuntime()))
print('Memory (RSS): ' + str(obj.getMemoryRSS()))
print('Memory (USS): ' + str(obj.getMemoryUSS()))
```

8.4 Geo-referenced Periodic-Frequent Pattern

8.4.1 *The Basic Model*

Continuing with the basic spatial database model, we denote a time series database, TSD, as a set of events. Each event represents $timestamp$ and $items$. That is, $TSD = \cup_{ts \in \mathbb{R}^+} \cup_{j=1}^{n} (ts, i_j)$, where $ts \in \mathbb{R}^+$ represents the timestamp. For brevity, the time series database can also represented by grouping the events by a timestamp as follows: A (irregular) **time series database** TSD is a collection of transactions. That is, $TSD = \{t_k, t_l, \cdots, t_m\}$, $k \leq l \leq m \leq |TSD|$, where $t_m = (ts, Y)$, where $Y \subseteq I$ is a pattern and $|TSD|$ represents the size of database. If a pattern $X \subseteq Y$, it is said that X occurs in transaction t_m. The timestamp of this transaction is denoted as ts_m^X. Let $TS^X = \{ts_k^X, ts_l^X, \cdots, ts_m^X\}$, $k,\ l,\ m \in (1,\ |TSD|)$, denote the set of all timestamps in which the pattern X has appeared in the database.

The number of transactions containing X in TSD is defined as the **support** of X and denoted as $sup(X)$. That is, $sup(X) = |TS^X|$. The pattern X is said to be a **frequent pattern** if the $sup(X) \geq minSup$, where $minSup$ refers to the user-specified *minimum support* value. Let ts_k^X and ts_l^X, $j \leq k < l \leq m$, be the two consecutive timestamps in TS^X. The time difference (or an inter-arrival time) between ts_l^X and ts_k^X is defined as a ***period*** of X, say p_a^X. That is, $p_a^X = ts_l^X - ts_k^X$. Let $P^X = (p_1^X, p_2^X, \cdots, p_b^X)$ be the set of all $periods$ for pattern X. The ***periodicity*** of X, denoted as $per(X) = max(p_1^X, p_2^X, \cdots, p_b^X)$. The frequent pattern X is said to be a **periodic-frequent pattern** if $per(X) \leq maxPer$, where $maxPer$ refers to the user-specified *maximum periodicity* value. The periodic-frequent pattern X is considered a GPFP if the maximum distance between its items is less than or equal to the user-specified *maximum distance* ($maxDist$) value. That is, X is a GPFP if $max(Dist(i_p, i_q)|\forall i_p, i_q \in X) \leq maxDist$, where $dist()$ is a distance function, say Euclidean distance, and $maxDist$ is a user-specified $maximum\ distance$ value [2].

Example 8.5 Let $I = \{p, q, r, s, t, u\}$ be a set of sensor identifiers (or items) in a network. The spatial locations of these items are shown in Table 8.5. A hypothetical time series database constituting these items is shown in Table 8.6. In the first transaction of Table 8.6, "1" represents the timestamp, and $\{p, q, r, s\}$ represents the transaction containing the items.[4] A similar statement can be made on remaining transactions in Table 8.6. The size of this temporal database, i.e., $|TSD| = 10$. The complete set of timestamps at which rs has occurred in Table 8.6, i.e., $TS^{rs} = \{1, 2, 5, 6, 9, 10\}$. The $support$ of "rs," i.e., $sup(rs) = |TS^{rs}| = |\{1, 2, 5, 6, 9, 10\}| = 6$. If the user-specified $minSup = 3$, then rs is said to be a frequent pattern because of $sup(rs) \geq minSup$. The periods for this pattern are: $p_1^{rs} = 1\ (= 1 - ts_{initial})$, $p_2^{rs} = 1\ (= 2 - 1)$, $p_3^{rs} = 3\ (= 5 - 2)$, $p_4^{rs} = 1\ (= 6 - 5)$, $p_5^{rs} = 3\ (= 9 - 6)$, $p_6^{rs} = 1\ (= 10 - 9)$, and $p_7^{rs} = 0\ (= ts_{final} - 10)$, where $ts_{initial} = 0$ represents the timestamp of initial transaction and $ts_{final} = |TSD| = 10$ represents the timestamp of final transaction in the database. The *periodicity* of rs, i.e., $per(rs) = maximum(1, 2, 3, 1, 3, 1, 0) = 3$. If the user-defined $maxPer = 4$, then the frequent pattern "rs" is said to be a periodic-frequent pattern because $per(rs) \leq maxPer$. The pattern rs is also a GPFP because $max(Dist(r, s)) \leq maxDist$.

Table 8.5 Location (or geo-referential) database

Item	Point	Item	Point
p	(2,3)	s	(2,3)
q	(6,8)	t	(1,5)
r	(1,4)	u	(3,4)

[4] A set of sensor identifiers in which pollution is very high at timestamp 1.

Table 8.6 Time series database. The items whose values were equal to 0 at a particular timestamp were removed for brevity

ts	Items	ts	Items
1	p, q, r, s	6	p, q, r, s
2	r, s, t	7	p, q
3	p, q, r, u	8	t, u
4	p, s, t	9	r, s
5	q, r, s, t, u	10	p, q, r, s, t, u

8.4.2 Handling the Search Space

The search space of geo-referenced periodic-frequent pattern mining is the same as that of geo-referenced frequent pattern mining. In other words, the search space of geo-referenced periodic-frequent patterns is $2^n - 1$, where n represents the total number of items in the data. One can effectively reduce the search space using the *anti-monotonic property* and the *neighborhood-aware depth-first search.*

8.4.3 Finding Geo-referenced Periodic-Frequent Patterns

The PAMI library provides GPFPMiner, PFS-ECLAT, and ST-ECLAT algorithms to find geo-referenced periodic-frequent patterns. Below is an example Python script demonstrating how to find the geo-referenced periodic-frequent patterns using the GPFPMiner algorithm.

Program 3: Finding Geo-referenced Periodic-Frequent Patterns

```
from PAMI.geoReferencedPeriodicFrequentPattern.basic import
↪  GPFPMiner  as alg

obj = alg.GPFPMiner("spatiotemporal_T10I4D100K.csv",
        "neighbors.txt", 1500, 500, '\t')

obj.mine()
obj.save('georeferencedPeriodicFrequentPatterns.txt')

geoperiodicFrequentPatternsDF= obj.getPatternsAsDataFrame()
print('Total No of patterns: ' +
↪  str(geoperiodicFrequentPatternsDF))
print('Runtime: ' + str(obj.getRuntime())) #measure the runtime

print('Memory (RSS): ' + str(obj.getMemoryRSS()))
print('Memory (USS): ' + str(obj.getMemoryUSS()))
```

8.5 Conclusion

This chapter explored the concepts and techniques for mining geo-referenced frequent patterns and geo-referenced periodic-frequent patterns. By incorporating both spatial and temporal dimensions, these patterns provide valuable insights into how data points (such as sensor readings) behave across time and space.

We began by defining geo-referenced frequent patterns, which consider spatial proximity between items. We then continued to define geo-referenced periodic-frequent patterns, which also take into account periodicity in addition to spatial proximity and frequency.

To manage the vast search space, we employed efficient search techniques, such as the anti-monotonic property and neighborhood-aware depth-first search, which significantly reduced the computational complexity. Additionally, we demonstrated how to use the PAMI library, which provides algorithms like FSP-growth, GPFP-Miner, and others for mining such patterns.

The Python script examples provided in the chapter show how these algorithms can be practically applied to datasets, illustrating the process of mining, saving, and analyzing geo-referenced patterns. These patterns are beneficial in real-world applications such as air pollution monitoring, environmental studies, and sensor networks, where understanding the interplay between spatial and temporal factors is crucial.

By identifying patterns that recur periodically and are geographically close, we can gain a deeper understanding of the dynamics within spatial networks and time series data, ultimately helping to improve decision-making and system optimization in various fields.

References

1. R. Uday Kiran, Sourabh Shrivastava, Philippe Fournier-Viger, Koji Zettsu, Masashi Toyoda, Masaru Kitsuregawa: Discovering Frequent Spatial Patterns in Very Large Spatiotemporal Databases. SIGSPATIAL/GIS 2020: 445–448.
2. Penugonda Ravikumar, R. Uday Kiran, Palla Likhitha, T. Chandrasekhar, Yutaka Watanobe, Koji Zettsu: Discovering Geo-referenced Periodic-Frequent Patterns in Geo-referenced Time Series Databases. DSAA 2022: 1-10

Chapter 9
Utility Databases: Representation, Creation, and Statistics

Abstract This chapter provides a comprehensive overview of utility databases, elucidating their theoretical foundations, practical applications, and significance in data mining and analysis. We begin with a formal definition of utility databases, detailing their structure and the identification of transactions using set theory. Practical considerations for storing and managing utility databases on computing devices are discussed, including formatting rules and transaction storage. Additionally, we explore methods for generating synthetic utility databases, which are crucial for testing and benchmarking algorithms in data mining. Techniques for converting structured dataframes into utility databases are also covered, expanding the scope of data analysis. Furthermore, we examine how to derive and interpret statistical details of utility databases to enhance understanding of their properties and optimize their use. By integrating theoretical insights with practical skills, this chapter provides users with the procedures to effectively manage, analyze, and leverage utility databases in diverse real-world applications, laying the groundwork for advanced data analysis.

9.1 Introduction

A structured certain nonbinary transactional database, or utility database [1], is an organized collection of transactions where items can take values from $(-\infty, \infty)$. A transactional identifier uniquely identifies each transaction. Utility databases are widely used in various real-world applications. For instance, in sensor networks, each transaction might represent the values recorded by sensors at specific time intervals. A utility database holds pixel data and their associated band values of satellite imagery. In social networks, utility databases track user interactions and activities over time, aiding in discovering trends and patterns in user behavior.

Figure 9.1 illustrates the critical factors in creating a utility database. The figure highlights the complex relationships and interactions essential for organizing and analyzing nonbinary data within the utility database framework. When all transactions in a utility database are accumulated over time, the result is a utility temporal database. Likewise, if the database contains spatial elements, it forms

U. K. Rage, *Hands-on Pattern Mining*,
https://doi.org/10.1007/978-981-96-6791-8_9

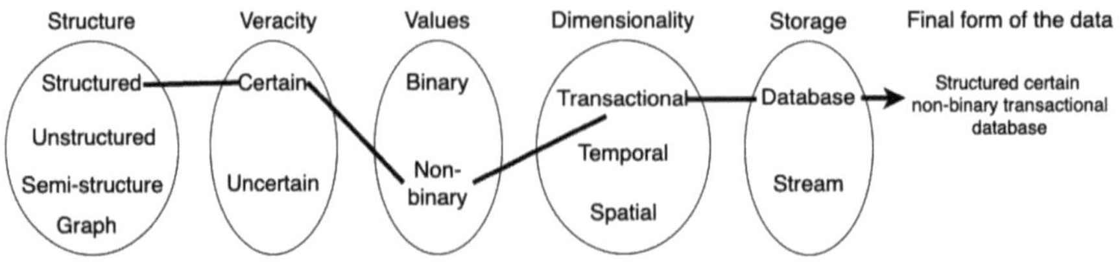

Fig. 9.1 Illustration of factors contributing to the creation of a utility database

a geo-referenced utility (transactional or temporal) database. For the sake of brevity, this chapter focuses on knowledge discovery within the conventional utility database.

This chapter covers the following key aspects of utility databases:

1. **Theoretical Representation**: Provides formal definitions of utility databases.
2. **Practical Representation**: Discusses how utility databases are structured, stored, and managed in computer systems, with examples of real-world applications.
3. **Synthetic Database Creation**: Covers techniques for generating synthetic databases, essential for testing, benchmarking, and evaluation purposes.
4. **Dataframe Conversion**: Explores methods for transforming structured dataframes into utility databases, expanding their use for comprehensive data analysis.
5. **Database Statistics**: Explains how to calculate and interpret statistical metrics to evaluate the characteristics and performance of these databases.

9.2 Theoretical Representation

A (transactional) utility database represents a collection of nonbinary transactions, each uniquely identified and containing a specific set of items and their values.

Definition 9.1 (External Utility Database) Let $J = \{j_1, j_2, \ldots, j_n\}$ where $n \geq 1$ denotes a collection of items. Each item $j_k \in J$ is associated with a positive value $eu(j_k)$, referred to as its **external utility**. This external utility indicates the significance of the item to the user. The **external utility database** (EUD) is defined as the aggregation of all items in J along with their respective external utility values, formally represented as

$$EUD = \{(j_1, eu(j_1)), (j_2, eu(j_2)), \ldots, (j_n, eu(j_n))\}.$$

Example 9.1 Let J = {Bread, Jam, Butter, Book, Pen} represent a collection of items available in a supermarket. The prices of these items, as illustrated in

Table 9.1a, constitute the external utility database. These external utility values also reflect the items' relative importance within the application context.

Definition 9.2 (Internal Utility Database) An **internal utility (transactional) database** is defined as a set of transactions $UDB = \{T_1, T_2, \ldots, T_m\}, m \geq 1$, where each transaction $T_i \in UDB$ is a subset of J and is uniquely identified by a positive integer $i \in \mathbb{Z}^+$, known as its transaction identifier (or tid). Within each transaction T_i, every item $j_k \in T_i$ is assigned a positive value $f(j_k, T_i)$, referred to as its **internal utility**. The internal utility typically represents the frequency of the item's occurrence within that transaction.

Example 9.2 A hypothetical internal utility database representing the purchases of items in J is presented in Table 9.1b. The first transaction indicates that a customer has purchased two units of Bread, one unit of Jam, and three units of Butter. Similar interpretations can be applied to the other transactions in this table.

Definition 9.3 (Utility Database) A utility database, denoted as UD, is an internal utility database in which the internal utility values of the items in a transaction are replaced by the product of their external and internal utility values.

Example 9.3 Table 9.1c displays the utility database generated by multiplying the internal and external utility values of all items in the transactions of an internal utility database.

Table 9.1 Hypothetical utility database of a supermarket

(a) External utility database

Item	Price ($Rs.$)
Bread	50
Jam	50
Butter	50
Book	20
Pen	20

(b) Internal utility database

tid	Items
1	(Bread,2), (Jam,1), (Butter,3)
2	(Bread,1), (Book, 2), (Pen,3)
3	(Jam,3), (Butter,1)
4	(Bread,2), (Jam,1), (Butter,2), (Pen,1)
5	(Book,3), (Pen,1)

(c) Utility database

tid	*Bread*	*Jam*	*Butter*	*Book*	*Pen*
1	100 ($= 2 \times 50$)	50 ($= 1 \times 50$)	150 ($= 3 \times 50$)	0 ($= 0 \times 20$)	0 ($= 0 \times 20$)
2	50	0	0	40	60
3	0	150	50	0	0
4	100	50	100	0	20
5	0	0	0	60	20

9.3 Practical Representation

A utility database is typically stored as a file on a computer. To effectively create and manage this file, the following rules should be observed:

- **One Transaction per Line**: Each line in the file corresponds to a single transaction. The line number implicitly serves as the transaction identifier (tid), so it is not explicitly stored in the file, which helps save space and reduce processing costs.
- **Three Components of a Transaction**: Each transaction consists of three components. The first component lists the items involved in the transaction. The second component presents the sum of the utility values of all items in that transaction. The final component contains the individual utility values for each item in the respective transaction.
- **Separator for the Components**: A colon delimiter must separate the three components of each transaction. Users cannot alter this delimiter.
- **Separator for the Elements in a Component**: The elements within a component can be separated by any delimiter, such as a tab, space, or comma. In the algorithms used in PAMI, a tab is considered the default delimiter for items or utility values within a component.

  Overall, the format of a transaction in a utility database is

$$item_1\langle sep\rangle item_2\langle sep\rangle \cdots : totalUtility : utility_1\langle sep\rangle utility_2\langle sep\rangle \cdots$$

Example 9.4 If the delimiter is a `tab`, the utility database shown in Table 9.1c would appear as follows:

```
Bread	Jam Butter:300:100	50	150
Bread	Book	Pen:150:50	40	60
Jam Butter:200:150	50
Bread	Jam Butter	Pen:270:100	50	100 20
Book	Pen:80:60	20
```

Important

The “colon” is the default separator used to divide the components of a transaction, and users cannot change this separator.

Important

The “tab” is the default separator used to split items or values within the component of a transaction, but users can modify this separator.

9.4 Creating Synthetic Utility Databases

The PAMI package provides a powerful and flexible tool for generating synthetic utility databases, which can be tailored to meet various requirements. This capability is precious for testing and developing algorithms in data mining and related fields. Users can customize the database based on their specific needs, including the number of transactions, the total number of items, average transaction length, and several other utility parameters.

To illustrate the process of creating a synthetic utility database, consider the following sample code. This example generates a database with 100,000 transactions, each containing an average of 10 items selected from a set of 1000 possible items:

Program 1: Generating a Synthetic Utility Database

```
from PAMI.extras.syntheticDataGenerator import UtilityDatabase as db
obj = db.UtilityDatabase(databaseSize=100000, avgItemsPerTransaction=10, numItems=1000, minInternalUtilityValue=1, maxInternalUtilityValue=100, minExternalUtilityValue=100, maxExternalUtilityValue=1000, sep='\t')
obj.create()
obj.save('utilityDatabase.csv')
utilityDataFrame = obj.getTransactions()
print('Runtime: ' + str(obj.getRuntime()))
print('Memory (RSS): ' + str(obj.getMemoryRSS()))
print('Memory (USS): ' + str(obj.getMemoryUSS()))
```

9.5 Deriving a Utility Database from a Dataframe

The PAMI package enables users to convert a dataframe into a utility database, which is ideal for transaction-based data analysis. Below is a Python code snippet illustrating how to use PAMI for this conversion:

Program 2: Converting a Dataframe into a Utility Database

```
from PAMI.extras.convert import DF2DB as alg
import pandas as pd
```

```
import numpy as np
data = np.random.randint(1, 100, size=(4, 4))
dataFrame = pd.DataFrame(data_4x4, columns=['Item1', 'Item2',
↪ 'Item3', 'Item4'])
obj = alg.DF2DB(dataFrame)
obj.convert2UtilityDatabase(oFile='utilityDB.csv')
print('Runtime: ' + str(obj.getRuntime()))
print('Memory (RSS): ' + str(obj.getMemoryRSS()))
print('Memory (USS): ' + str(obj.getMemoryUSS()))
```

9.6 Understanding the Statistical Details

The `dbStats` sub-sub-package in the `extras` sub-package of PAMI provides users statistical details about a utility database. This functionality is essential for understanding the properties and characteristics of the database, which can be crucial for various data analysis tasks. The statistical details provided by dbStats include:

1. Database size
2. Total number of items in a database
3. Minimum, average, and maximum lengths of the transactions
4. Minimum, average, and maximum utility value of a transaction
5. Standard deviation of transactional sizes
6. Variance in transaction sizes
7. Sparsity
8. Frequencies of the items
9. Distribution of transactional lengths
10. Distribution of items' utility values

Here is an example of how to use the `dbStats` to obtain the statistics:

Program 3: Deriving the Statistical Details of Utility Database

```
from PAMI.extras.dbStats import UtilityDatabase as stat

obj = stat.UtilityDatabase("utilityDatabase.csv")
obj.run()
obj.printStats()
obj.plotGraphs()
```

9.7 Variants of Utility Databases

9.7.1 Temporal Utility Database

A temporal utility database [2] generalizes the basic (transactional) utility database by ordering the transactions with respect to a timestamp. The time gap between all the transactions remains constant in the utility database; we call that database a uniform temporal utility database. If the time gap between the transactions varies in the utility database, we call that database a nonuniform temporal utility database. One can convert any transactional utility database into a temporal utility database by simply concatenating the timestamp at the beginning of the transaction.

Overall, the format of a transaction in a temporal utility database is

$$timestamp\langle sep\rangle item_1\langle sep\rangle \cdots : totalUtility : utility_1\langle sep\rangle \cdots$$

Example 9.5 If the delimiter is a `tab`, a utility database shown in Table 9.1c can be converted into a temporal utility database as follows:

```
1    Bread    Jam Butter:300:100   50   150
2    Bread    Book     Pen:150:50   40   60
3    Jam Butter:200:150   50
4    Bread    Jam Butter   Pen:270:100    50   100 20
5    Book     Pen:80:60   20
```

9.7.2 Geo-referenced Transactional Utility Database

A geo-referenced transactional utility database [3] contains items with spatial information, such as points, lines, and polygons. Overall, the format of a transaction in a geo-referenced transactional utility database is

$coordinates_1\langle sep\rangle coordinates_2\langle sep\rangle coordinates_3\langle sep\rangle \cdots$: $totalUtility$: $utility_1\langle sep\rangle utility_2\langle sep\rangle utility_3\langle sep\rangle \cdots$

A sample geo-referenced transactional utility database with a `tab` delimiter is

```
POINT(0 1)  POINT(2 1)  POINT(1 0):4:1  2   1
POINT(0 1)  POINT(1 0)  POINT(1 2):10:2 6   2
POINT(2 1)  POINT(1 2)  POINT(1 1)  POINT (1 5):8:1 4   2   1
POINT(0 1)  POINT(1 0)  POINT(1 1)  POINT (1 5):10:1 6  1  1
POINT(0 1)  POINT(1 1)  POINT(1 5)  POINT(1 6):20:6 6   10  2
POINT(0 1)  POINT(1 0)  POINT(1 1)  POINT(1 5):15:1 1   6   6
POINT(0 1)  POINT(1 0)  POINT(1 1)  POINT(1 5):6:2  2   1   1
```

9.7.3 Geo-referenced Temporal Utility Database

A geo-referenced temporal utility database is a temporal utility database containing items with spatial information, such as points, lines, and polygons. Overall, the format of a transaction in a geo-referenced temporal utility database is

$timestamp\langle sep\rangle coordinates_1\langle sep\rangle coordinates_2\langle sep\rangle coordinates_3\langle sep\rangle \cdots : totalUtility : utility_1\langle sep\rangle utility_2\langle sep\rangle utility_3\langle sep\rangle \cdots$

A sample geo-referenced temporal utility database with a `tab` delimiter is

```
1 POINT(0 1)  POINT(2 1)  POINT(1 0):4:1  2   1
2 POINT(0 1)  POINT(1 0)  POINT(1 2):10:2 6   2
3 POINT(2 1)  POINT(1 2)  POINT(1 1)  POINT(1 5):8:1 4  2  1
4 POINT(0 1)  POINT(1 0)  POINT(1 1)  POINT(1 5):10:1 6  1  1
5 POINT(0 1)  POINT(1 1)  POINT(1 5)  POINT(1 6):20:6 6  10  2
6 POINT(0 1)  POINT(1 0)  POINT(1 1)  POINT(1 5):15:1 1  6  6
7 POINT(0 1)  POINT(1 0)  POINT(1 1)  POINT(1 5):6:2  2  1  1
```

9.8 Conclusion

This chapter has provided a comprehensive overview of utility databases, covering both their theoretical foundations and practical applications. We started with a formal definition of utility databases, explaining how transactions are structured and identified through set theory. Next, we examined the practical aspects of storing and managing these databases on computing devices, including the rules for formatting and storing transactions. We also explored methods for generating synthetic utility databases, essential for testing and benchmarking data mining algorithms. Additionally, we discussed techniques for converting structured dataframes into utility databases, broadening the data analysis scope. Finally, we analyzed how to derive and interpret statistical details of utility databases, enhancing our understanding of their properties and optimizing their usage.

References

1. R. Uday Kiran, T. Yashwanth Reddy, Philippe Fournier-Viger, Masashi Toyoda, P. Krishna Reddy, Masaru Kitsuregawa: Efficiently Finding High Utility-Frequent Itemsets Using Cutoff and Suffix Utility. PAKDD (2) 2019: 191–203.
2. Pradeep Pallikila, Pamalla Veena, R. Uday Kiran, Ram Avatar, Sadanori Ito, Koji Zettsu, P. Krishna Reddy: Discovering Top-k Spatial High Utility Itemsets in Very Large Quantitative Spatiotemporal databases. IEEE BigData 2021: 4925-4935
3. Sai Chithra Bommisetty, Penugonda Ravikumar, Rage Uday Kiran, Minh-Son Dao, Koji Zettsu: Discovering Spatial High Utility Itemsets in High-Dimensional Spatiotemporal Databases. IEA/AIE (1) 2021: 53-65

Chapter 10
Pattern Discovery in Utility Databases

Abstract This chapter explores the analytical process of mining high utility patterns from utility databases, emphasizing the significance and extraction of meaningful patterns based on their utility. It introduces key concepts, such as utility calculation for items and patterns within transactions, and discusses the challenge of large search space in pattern mining. Techniques like the EFIM algorithm are highlighted to discover high utility patterns efficiently. Additionally, the chapter extends to high utility frequent pattern mining, which integrates both utility and support constraints to refine pattern discovery by excluding infrequent but high utility patterns. The HUIM algorithm is also demonstrated with a practical Python implementation, providing a robust framework for mining utility databases and uncovering critical insights for real-world applications.

10.1 Introduction

The previous chapter provided a comprehensive overview of utility databases, covering their construction, practical representation, and methods for deriving statistical insights. This chapter focuses on the analytical dimension, extracting and analyzing meaningful patterns within a utility database.

This chapter delves into the following topics:

1. **High utility pattern discovery**: We will define the notion of high utility patterns that might exist in a utility database.
2. **High utility frequent pattern discovery**: This subsection describes the model of finding high utility frequent patterns in a utility database.

Chapter 9 introduced the foundational concepts of utility databases, including key terms such as "pattern," "transaction," and "utility database." We will continue using these terms consistently throughout this chapter to streamline the discussion and minimize redundancy. For readers who may have missed the previous chapter, we recommend reviewing at least Sect. 9.2 to familiarize themselves with the essential concepts and terminologies.

U. K. Rage, *Hands-on Pattern Mining*,
https://doi.org/10.1007/978-981-96-6791-8_10

10.2 High Utility Patterns

High utility patterns [1] are an essential class of regularities that can be identified within the utility databases. This section delves into finding high utility patterns in detail, emphasizing their importance for effectively mining and analyzing patterns related to user preferences and activities. Knowledge of high utility patterns is essential for uncovering critical relationships within the data and serves as the basis for more advanced pattern mining techniques.

10.2.1 Basic Model

We introduce the model of high utility patterns based on the terminology of the utility database described in Sect. 9.2. Examples are illustrated using the data in Table 9.1.

Definition 10.1 (Utility of an Item in a Transaction) The utility of an item j_k in a transaction T_i, denoted as $u(j_k, T_i)$, represents the product of its external and internal utility values. That is, $u(j_k, T_i) = eu(j_k) \times f(j_k, T_i)$.

Example 10.1 Continuing with the previous example, the *utility* (or income) of an item *Bread* in the first transaction, i.e., $u(Bread, T_1) = eu(Bread) \times f(Bread, T_1) = 50 \times 2 = 100\ Rs$.

Definition 10.2 (Utility of a Pattern in a Transaction) The utility of a pattern X in a transaction T_i is denoted as $u(X, T_i) = \Sigma_{j_k \in X} u(j_k, T_i)$ if $X \subseteq T_i$.

Example 10.2 The set of items "*Bread*" and "*Jam*," i.e., $\{Bread, Jam\}$, is a pattern. The utility (or *income*) of $\{Bread, Jam\}$ in T_1, $u(\{Bread, Jam\}, T_1) = u(Bread, T_1) + u(Jam, T_1) = 100 + 50 = 150\ Rs$.

Definition 10.3 (Utility of a Pattern in a Database) The utility of a pattern X in the database UD, denoted as $u(X) = \Sigma_{T_i \in g(X)} u(X, T_i)$, where $g(X)$, is the set of transactions containing X.

Example 10.3 In Table 9.1, $\{Bread, Jam\}$ has appeared in the transactions whose *tids* are 1 and 4. Therefore, $g(\{Bread, Jam\}) = \{T_1, T_4\}$. The *utility* (or *income*) of $\{Bread, Jam\}$ in each of these transactions, i.e., $u(\{Bread, Jam\}, T_1) = 150$ and $u(\{Bread, Jam\}, T_4) = 150$. Therefore, the utility (or *income*) of $\{Bread, Jam\}$ in the entire database, i.e., $u(\{Bread, Jam\}) = 150 + 150 = 300\ Rs$.

Definition 10.4 (High Utility Pattern) A pattern X is a high utility pattern if its $u(X) \geq minUtil$, where $minUtil$ represents the user-specified minimum utility value. A high utility pattern X is expressed as $X\ [utility = u(X)]$.

Example 10.4 If the user-specified $minUtil = 250$, then the pattern $\{Bread, Jam\}$ is a high utility pattern because $u(\{Bread, Jam\}) \geq minUtil$. This pattern is expressed as $\{Bread, Jam\}$ $[utility = 300]$.

10.2.2 Search Space

The space of items in a utility database raises an itemset lattice. This lattice represents the search space of high utility pattern mining. Thus, the search space size is $2^n - 1$, where n represents the total number of items in a database. This vast search space followed the inability to employ the *Apriori property* to reduce the search space, making the high utility pattern mining a computationally expensive task. To make high utility pattern mining practicable on huge databases, its mining algorithms employ different upper-bound utility measures, such as *total utility*, *remaining utility*, and *local utility*, to reduce the search space considerably.

10.2.3 Finding High Utility Patterns

Several algorithms, such as EFIM, HMiner, and UPGrowth, were described in the literature to find high utility patterns. Although no universally acceptable best algorithm exists for finding these patterns in any utility database, most researchers utilize the EFIM as it was generally found to be faster than the other algorithms. Below is a sample Python script for finding high utility patterns using the EFIM algorithm available in the PAMI package.

Program 1: High Utility Pattern Discovery Using EFIM

```
from PAMI.highUtilityPattern.basic import EFIM  as alg
obj = alg.EFIM(iFile='Utility_T10I4D100K.csv',  minUtil=10000,
↪  sep='\t')
obj.mine()              #start the mining process
obj.save('utilityPatterns.txt') #save the patterns
utilityPatternsDF= obj.getPatternsAsDataFrame()
print('# patterns: ' + str(len(utilityPatternsDF)))
print('Runtime: ' + str(obj.getRuntime()))
print('Memory (RSS): ' + str(obj.getMemoryRSS()))
print('Memory (USS): ' + str(obj.getMemoryUSS()))
```

10.3 High Utility Frequent Patterns

Since the basic model of high utility patterns determines a pattern's interestingness without considering its *support* within the data, uninteresting patterns with very low *support* may be generated as high utility patterns. Many high utility patterns were uninteresting, often appearing infrequently in the data. To prune the high utility patterns that have appeared infrequently in the data, the researchers extended the basic model of high utility pattern mining to find high utility frequent patterns [2] by considering additional constraints, namely *minimum support* ($minSup$). We now describe the extended model of high utility frequent patterns and describe the process of finding them.

10.3.1 Basic Model

Definition 10.5 (Support of a Pattern) Let $P \subseteq J$ be a pattern. The *support* of P in a utility database UD is defined as

$$\text{sup}(P) = \frac{\text{freq}(P)}{|UD|},$$

where freq(P) denotes the frequency of pattern P in UD, and $|UD|$ represents the total number of transactions in the database.

Example 10.5 In Table 9.1, the high utility pattern $\{Bread, Jam\}$ has appeared in the transactions whose $tids$ are 1 and 4. Thus, the frequency of $\{Bread, Jam\}$ is 2. The $support$ of $\{Bread, Jam\}$, i.e., $sup(\{Bread, Jam\}) = \frac{2}{5} = 0.4(= 40\%)$.

Definition 10.6 (High Utility Frequent Pattern) A high utility pattern P is considered to be a high utility frequent pattern if $s(P) \geq minSup$, where $minSup$ represents the user-specified minimum support. A high utility frequent itemset P is expressed as $P\ [support = s(P),\ utility = u(P)]$.

Example 10.6 If $minSup = 0.3$, then the high utility pattern $\{Bread, Jam\}$ is said to be a high utility frequent pattern because $s(\{Bread, Jam\}) \geq minSup$. This pattern is expressed as $\{Bread, Jam\}\ [support = 0.4,\ utility = 300]$.

10.3.2 Search Space

The search space of high utility frequent pattern mining is $2^n - 1$, the same as that of the high utility pattern mining. We can effectively reduce the search space using the anti-monotonic property of the *support* measure. Overall, high utility frequent pattern mining is computationally less expensive than the high utility pattern mining.

10.3.3 Finding High Utility Frequent Patterns

An efficient algorithm, namely high utility itemset mining (HUIM), has been described in the literature to find high utility frequent patterns. Below is a sample Python script for finding high utility frequent patterns using the HUIM algorithm available in the PAMI package.

Program 2: High Utility Frequent Pattern Discovery Using HUIM

```
from PAMI.highUtilityFrequentPattern.basic import HUFIM  as alg
obj = alg.HUFIM(iFile='Utility_T10I4D100K.csv', minUtil=10000,
↪  minSup=500, sep='\t')
obj.mine()
obj.save('utilityFrequentPatternsAtMinSup.txt')
utilityFPDF= obj.getPatternsAsDataFrame()
print('Total No of patterns: ' + str(len(utilityFPDF)))
print('Runtime: ' + str(obj.getRuntime()))
print('Memory (RSS): ' + str(obj.getMemoryRSS()))
print('Memory (USS): ' + str(obj.getMemoryUSS()))
```

10.4 Conclusion

In this chapter, we explored the analytical dimension of utility databases by focusing on high utility pattern discovery and its various extensions. High utility patterns are essential for uncovering significant relationships within data, enabling the extraction of patterns based on their utility rather than their frequency alone. We introduced the basic model for high utility patterns, highlighting how the utility of items and patterns is calculated in individual transactions and across the entire database.

We discussed the importance of employing upper-bound utility measures to reduce computational complexity and demonstrated how the EFIM algorithm can efficiently find high utility patterns. In the second part of the chapter, we explored the concept of high utility frequent patterns, an extension that incorporates both utility and frequency constraints. This approach allows for the discovery of valuable and frequent patterns, thus refining the results by eliminating infrequent patterns that may not be of practical significance. The chapter concluded with a Python implementation of the HUIM algorithm to find high utility frequent patterns.

References

1. Cheng-Wei Wu, Philippe Fournier-Viger, Philip S. Yu, Vincent S. Tseng: Efficient Mining of a Concise and Lossless Representation of High Utility Itemsets. ICDM 2011: 824-833
2. R. Uday Kiran, T. Yashwanth Reddy, Philippe Fournier-Viger, Masashi Toyoda, P. Krishna Reddy, Masaru Kitsuregawa: Efficiently Finding High Utility-Frequent Itemsets Using Cutoff and Suffix Utility. PAKDD (2) 2019: 191–203.

Chapter 11
Sequence Databases: Representation, Creation, and Statistics

Abstract Sequence databases, an extension of transactional databases, store ordered collections of transactions, making them invaluable for applications in healthcare, e-commerce, and web analytics. These databases structure transactions sequentially, often based on time or customer behavior, to reveal patterns that can drive socioeconomic development. This chapter introduces sequence databases by defining their mathematical representation through set theory, followed by an exploration of practical storage and implementation techniques. It details methods for generating synthetic sequence databases, which facilitate benchmarking and algorithm testing, and explains how to convert dataframes into sequential databases for broader analysis. Additionally, the chapter introduces statistical procedures for extracting critical insights, such as item frequencies and sequence length variations, from sequence databases. By combining theoretical foundations with practical applications, this chapter equips readers with essential tools for managing and analyzing sequential data, setting the stage for advanced data mining and analysis techniques.

11.1 Introduction

A structured certain binary sequential database, or simply a sequence database, is a variant of a transactional database, where transactions are grouped and ordered based on a metric, say $timestamp$ or $customer\ identifier$. Many real-world applications, such as healthcare, weblogs, and e-commerce, naturally produce transactional databases that can be represented as a sequence database. Useful information that can empower the end users to achieve socioeconomic development lies hidden in this data.

This chapter covers the following key aspects of sequence databases:

1. **Theoretical Representation**: The formal definition of a sequence database using set theory
2. **Practical Representation**: How computer systems implement and store sequence databases

U. K. Rage, *Hands-on Pattern Mining*,
https://doi.org/10.1007/978-981-96-6791-8_11

3. **Synthetic Database Creation**: Techniques for generating synthetic sequence databases for testing and benchmarking
4. **Dataframe Conversion**: Methods to convert structured dataframes into sequential databases for broader data analysis applications
5. **Database Statistics**: How to derive statistical details about a sequence database

11.2 Theoretical Representation

A sequence represents an ordered collection of transactions. A sequential database [1] represents a collection of sequences. Formally:

Let $O = \{o_1, o_2, \cdots, o_n\}$, $n \geq 1$, be a set of items (or objects). Let $G \subseteq O$ be a pattern (or an itemset). A pattern containing k number of items is a k-pattern. Let $|G|$ denote the cardinality of a pattern, i.e., $|G| = k$. Without loss of generality, let us assume there exists a total order on objects $\succ$, say lexicographical order. A *sequence*, denoted as s_a, $a \geq 1$, is an ordered list of itemsets. That is, $s_a = \{G_1, G_2, \cdots, G_p\}$, $p \geq 1$. A sequence s_a is said to be an α-sequence if it contains a α number of items, i.e., $\alpha = |G_x| \forall G_x \in s_a$. A sequence database, denoted as SDB, is a list of sequences. That is, $SDB = \{s_1, s_2, \cdots, s_q\} = \cup_{sid=1}^{q} s_{sid}$, where $q \geq 1$ and $sid \geq 1$ denotes the *sequence identifier*. The size (or $cardinality$) of a sequence database, denoted as $|SDB| = q$, where q represents the total number of sequences in a database.

Example 11.1 Let $O = \{a, b, c, d, e, f, g\}$ be a set of items. The set of items a and b, i.e., $\{a, b\}$ (or ab, in short), is a pattern. This pattern contains two items. Henceforth, it is a 2-pattern with the cardinality of 2 (= $|ab|$). Let $a \succ b \succ b \succ c \succ d \succ e \succ f \succ g$ be the lexicographical order of items. A sequence $s_1 = \langle\ ab,\ c,\ ef\ \rangle$ represents the sequential occurrence order of three patterns. This sequence contains five distinct items. Hence, it is a 5-sequence. A hypothetical hourly sales transactional database containing these five items is shown in Table 11.1a. This database can be shown as a sequence database as in Table 11.1b. This sequence database contains four sequences. Henceforth, the size of SDB, i.e., $|SDB| = 4$.

11.3 Practical Representation

A sequential database is usually stored as a file on a computer. To properly create and manage this file, follow these rules:

- **One Sequence per Line**: Each line in the file represents a single sequence. The line number implicitly acts as the sequence identifier (sid), so it is not explicitly stored in the file to save space and reduce processing costs.

Table 11.1 Hypothetical transactional database of a supermarket

(a) Hypothetical hourly sales data

Hour	TID	Items	Hour	TID	Items
1	1	*ab*	3	8	*d*
1	2	*c*	3	9	*abde*
1	3	*def*	3	10	*ac*
2	4	*ad*	4	11	*ab*
2	5	*cd*	4	12	*c*
2	6	*af*	4	13	*aeg*
3	7	*bc*	4	14	*df*

(b) Sequence database

sid	Sequences
1	$\langle ab, c, def \rangle$
2	$\langle ad, cd, af \rangle$
3	$\langle bc, d, abde, ac \rangle$
4	$\langle ab, c, aeg, df \rangle$

- **Patterns Separated by a Delimiter**: The patterns in a sequence are separated by a colon. Users cannot overwrite this delimiter.
- **Unique Items per Pattern**: Each item should appear only once within a pattern. However, an item can appear any number of times in a sequence.
- **Items Separated by a Delimiter in a Pattern**: Items in a pattern are separated by a delimiter, such as a space or tab. The PAMI algorithms use a `tab` as the default delimiter, but users can choose other delimiters like commas or spaces.

Overall, the format of a sequence in a sequential is

$$item_1 \langle sep \rangle item_2 \langle sep \rangle item_3 : item_1 \langle sep \rangle item_2 : \cdots$$

Example 11.2 If the delimiter is a `tab`, the sequential database shown in Table 11.1b would look like this:

```
a   b:c:d   e   f
a   d:c d:a f
b   c:d:a   b   d   e:a c
a   b:c:a   e   g:d f
```

11.4 Creating Synthetic Sequence Databases

The PAMI package offers a powerful and flexible tool for generating synthetic sequential databases tailored to various needs. This capability is invaluable for testing and developing algorithms in data mining and related fields. Users can customize the database to suit their specific requirements, including the number of transactions, the total number of items, and the average transaction length.

To illustrate the creation of a synthetic sequential database, consider the following sample code.

Program 1: Generating Synthetic Sequential Database

```
from PAMI.extras.syntheticDataGenerator import SequentialDatabase as db
obj = db.SequentialDatabase( databaseSize=100000, avgItemsPerPatterns=10, avgPatternsPerSequence=10, numItems=1000, sep='\t')
obj.create()
obj.save('sequentialDatabase.csv')
#read the generated sequences into a dataframe
sequentialDataFrame=obj.getSequences()
#stats
print('Runtime: ' + str(obj.getRuntime()))
print('Memory (RSS): ' + str(obj.getMemoryRSS()))
print('Memory (USS): ' + str(obj.getMemoryUSS()))
```

11.5 Deriving a Sequence Database from a Dataframe

The PAMI package enables users to convert a dataframe into a sequence database, which is ideal for transaction-based data analysis. Below is a Python code snippet illustrating how to use PAMI for this conversion:

Program 2: Converting a Dataframe into a Utility Database

```
from PAMI.extras.convert import DF2DB as alg
import pandas as pd
import numpy as np
data = np.random.randint(1, 100, size=(4, 4))
dataFrame = pd.DataFrame(data, columns=['Item1', 'Item2', 'Item3', 'Item4'])

customerID= np.random.randint(1, 3, size=(4, 1))
customerIDdataFrame = pd.DataFrame(customerID, columns=['customerID'])

 dataFrame = pd.concat([customerIDdataFrame, dataFrame], axis=1)

obj = alg.DF2DB(dataFrame)
```

```
obj.convert2SequenceDatabase(oFile='sequentialDatabase.csv',
↪   condition='>=', value=20)
print('Runtime: ' + str(obj.getRuntime()))
print('Memory (RSS): ' + str(obj.getMemoryRSS()))
print('Memory (USS): ' + str(obj.getMemoryUSS()))
```

11.6 Knowing the Statistical Details

The `stats` sub-sub-package in the `extras` sub-package of PAMI provides users with statistical details about a sequential database. This functionality is essential for understanding the properties and characteristics of the database, which can be crucial for various data analysis tasks. The statistical details provided by `stats` include:

1. Database size
2. Total number of items in a database
3. Minimum, average, and maximum lengths of the sequences
4. Standard deviation of sequence sizes
5. Variance in sequence sizes
6. Sparsity
7. Frequencies of the items
8. Distribution of transactional lengths

Here is an example of how to use the `dbStats` to obtain the statistics:

Program 2: Deriving the Statistical Details

```
from PAMI.extras.stats import SequentialDatabase as stat

obj = stat.SequentialDatabase("sequentialDatabase.csv")
obj.run()
obj.printStats()

obj.plotGraphs()
```

11.7 Conclusion

This chapter has provided a comprehensive overview of sequence databases, from their theoretical underpinnings to practical applications. We began with a formal definition of sequence databases, detailing how sequences are structured and identified using set theory. We then explored the practical aspects of how these databases are stored and managed on computing devices, including the rules for formatting and storing transactions.

We also discussed methods for generating synthetic sequence databases, which are crucial for testing and benchmarking data mining algorithms. Finally, we examined how to derive and interpret statistical details of sequential databases to better understand their properties and optimize their usage.

Understanding these concepts and techniques equips users with the tools to manage, analyze, and leverage sequential databases in various real-world applications. The combination of theoretical knowledge and practical skills discussed here lays the foundation for advanced data analysis.

References

1. Wensheng Gan, Jerry Chun-Wei Lin, Philippe Fournier-Viger, Han-Chieh Chao, Philip S. Yu: A Survey of Parallel Sequential Pattern Mining. ACM Trans. Knowl. Discov. Data 13(3): 25:1–25:34 (2019)

Chapter 12
Pattern Discovery in Sequence Databases

Abstract Sequential pattern mining is a powerful analytical tool used to uncover significant patterns within ordered data, enabling insights into recurring trends and behaviors. This chapter delves into the discovery of frequent sequence patterns in sequential databases, a process valuable across domains such as e-commerce, bioinformatics, and web usage analysis. We begin with foundational definitions and introduce the concept of sequence support as a measure of pattern significance. Leveraging the minimum support constraint, we discuss strategies to reduce search space and examine the well-known GSP algorithm to facilitate efficient pattern discovery. A practical implementation using the GSP algorithm offers insights into memory and runtime considerations critical for large datasets. This chapter equips readers with the algorithms needed to perform effective sequence pattern mining by combining theoretical foundations with practical applications, thereby enhancing data-driven decision-making in complex sequential datasets.

12.1 Introduction

The previous chapter provided a comprehensive overview of sequential databases, covering their construction, practical representation, and methods for deriving statistical insights. This chapter focuses on the analytical dimension, extracting and analyzing meaningful patterns, especially frequency sequence patterns, within sequential data.

Chapter 10 introduced the foundational concepts of sequential databases, including key terms such as "pattern," "sequence," and "sequential database." We will continue using these terms consistently throughout this chapter to streamline the discussion and minimize redundancy. For readers who may have missed the previous chapter, we recommend reviewing at least Sect. 11.2 to familiarize themselves with the essential concepts and terminologies.

U. K. Rage, *Hands-on Pattern Mining*,
https://doi.org/10.1007/978-981-96-6791-8_12

12.2 Frequent Sequence Patterns

12.2.1 Basic Model

Definition 12.1 Let $s_p = \langle A_1, A_2, \cdots, A_u \rangle$ and $s_q = \langle B_1, B_2, \cdots B_u \cdots B_v \rangle$, where $p \neq q$ and $1 \leq u \leq v$. We say that s_p is contained (or occurs) in S_q, i.e., $s_p \sqsubseteq s_q$, if and only if there exist integers $1 \leq h_1 \leq h_2 \leq \cdots \leq h_u \leq v$ such that $A_1 \subseteq B_{h_1}, A_2 \subseteq B_{h_2}, \cdots, A_u \subseteq B_{h_u}$.

Example 12.1 Let $s_x = \langle ab, c \rangle$ be a sequence. This sequence is contained in $\langle ab, c, def \rangle$, which is s_1 in Table 11.1b. Henceforth, s_x is a subsequence of s_1, denoted as $s_x \sqsubseteq s_1$. We can also state s_x occurs in s_1 for simplicity purpose.

Definition 12.2 (The Support of a Sequence) The *support* of a sequence s_p in a sequence database SDB is defined as the number of sequences that contain s_p, and is denoted by $sup(s_p)$. That is, $sup(s_p) = |s| s \sqsubseteq s_p \wedge s \in SDB|$. Please note that the *support* of a sequence can also be expressed in percentage of $|SDB|$.

Example 12.2 Continuing the previous example, the sequence s_x occurs in s_1 and s_4 of Table 11.1b. Henceforth, the *support* of s_x, i.e., $sup(s_x) = |\{s_1, s_4\}| = 2$.

Definition 12.3 (A Frequent Sequence Pattern [1]) A sequence s_p is said to a frequent sequence pattern if $sup(s_p) \geq minSup$, where $minSup$ represents the user-specified *minimum support* value.

Example 12.3 If the user-specified $minSup = 2$, then the sequence s_x is said to be a frequent sequence pattern as $sup(s_x) \geq minSup$. Similarly, another sequence, $s_y = \langle ab, d \rangle$, which occurs in s_1 and s_4 of Table 11.1b and has *support* of 2, is also said to be a frequent sequence pattern as $sup(s_y) \geq minSup$.

12.2.2 Search Space

The sequence lattice is derived from the space of itemsets and serves as the search space for mining sequential patterns. Here, the search space for sequence pattern mining is quantified as n^k, where k signifies the maximum sequence length and n denotes the aggregate count of items in the database. One can effectively reduce this colossal search space using the anti-monotonic property of the $minSup$ constraint.

12.2.3 Mining Algorithm

The literature describes several algorithms, such as PrefixSpan [2], SPAM [3], and SPADE [4], for finding frequent sequence patterns in the data. Although no universally acceptable best algorithm exists for finding these patterns in a sequential database, most researchers utilize the GSP algorithm, which was generally faster than the other algorithms. Below is a sample Python script for finding frequent sequence patterns using the GSP algorithm available in the PAMI package.

Program 1: Frequent Sequence Pattern Discovery Using GSpan

```
obj = alg.GSP('airDatabase.txt', minSup, '\t')
obj.mine()
# Retrieve discovered patterns and resource usage
Patterns = obj.getPatterns()
print("Total number of Frequent Sequence Patterns:",
len(Patterns))
# Display memory and runtime statistics
memUSS = obj.getMemoryUSS()
print("Total Memory in USS:", memUSS)
memRSS = obj.getMemoryRSS()
print("Total Memory in RSS", memRSS)
runTime = obj.getRuntime()
print("Total ExecutionTime in ms:", runTime)
```

12.3 Conclusion

This chapter examined the process of identifying frequent sequence patterns in sequential databases, highlighting their significance in fields like market analysis, web usage mining, and bioinformatics. We defined vital concepts such as sequence support and frequent patterns and explored methods to efficiently reduce search space using the minimum support constraint. We have also described a procedure to find the frequent sequence patterns using the GSP algorithm. This chapter provided theoretical insights and practical tools, equipping researchers and practitioners to leverage sequential pattern mining for meaningful insights and predictive analytics in complex data environments.

References

1. Wensheng Gan, Jerry Chun-Wei Lin, Philippe Fournier-Viger, Han-Chieh Chao, Philip S. Yu: A Survey of Parallel Sequential Pattern Mining. ACM Trans. Knowl. Discov. Data 13(3): 25:1–25:34 (2019)

2. Jian Pei, Jiawei Han, B. Mortazavi-Asl, Jianyong Wang, H. Pinto, Qiming Chen, U. Dayal, and Mei-Chun Hsu. 2004. Mining sequential patterns by pattern-growth: the PrefixSpan approach. IEEE TKDE 16, 11 (2004), 1424–1440.
3. Jay Ayres, Johannes Gehrke, Tomi Yiu, and Jason Flannick. 2002. Sequential PAttern Mining using A Bitmap Representation. ACM SIGKDD (07 2002).
4. Mohammed J Zaki. 2001. SPADE: An efficient algorithm for mining frequent sequences. Machine learning 42 (2001), 31–60.

Part II
Advanced Concepts

Chapter 13
Mining Symbolic Sequences

Abstract Symbolic sequence databases are widely used in fields such as bioinformatics, where analyzing DNA, RNA, and protein sequences is critical for understanding diseases and developing new drugs. This chapter presents an overview of symbolic sequence databases, focusing on their mathematical, practical representations and methods for generating and analyzing synthetic sequence data. We also explore techniques for discovering frequent contiguous patterns in symbolic sequences, essential for uncovering hidden relationships and insights within large datasets. The chapter introduces the PAMI library, which implements powerful tools such as the `PositionMining` algorithm for mining frequent contiguous patterns. These tools, along with database statistics and synthetic data generation capabilities, provide a comprehensive framework for researchers to analyze and extract meaningful patterns from symbolic sequence data.

13.1 Introduction

This chapter introduces symbolic sequence databases, which store continuous sequences of symbols or characters—data crucial for bioinformatics applications, such as analyzing DNA, RNA, and protein interactions to understand diseases and aid in drug development better. The key topics covered in this chapter are:

1. **Theoretical Representation**: Establishes a formal definition of a symbolic sequence database using set theory.
2. **Practical Representation**: Describes how sequence databases are implemented and stored in computer systems, focusing on the practical aspects of data handling.
3. **Synthetic Database Creation**: Explains methods for generating synthetic symbolic sequence databases, allowing researchers to test and benchmark algorithms in controlled environments.
4. **Database Statistics**: Outlines statistical approaches to derive insights from sequence databases, including details on symbolic sequence length distributions, data variation, and other statistical properties.

U. K. Rage, *Hands-on Pattern Mining*,
https://doi.org/10.1007/978-981-96-6791-8_13

5. **Pattern Discovery**: Details techniques for identifying frequent contiguous patterns within symbolic sequence data, essential for detecting biologically significant motifs or anomalies.

These sections offer theoretical foundations and practical tools, guiding readers through essential concepts and methods for working with symbolic sequence databases.

13.2 Theoretical Representation

A symbolic sequence is an ordered collection of symbols (or characters). Formally: let $\Sigma = \{A, B, \cdots, Z\}$ represent the set of symbols (also known as the alphabets). A sequence S is defined as an ordered arrangement of these symbols, expressed as $S = \langle s_1, s_2, \ldots, s_n \rangle$, where each $s_i \in \Sigma$, $1 \leq i \leq n$.

Example 13.1 Let $\Sigma = \{A, C, G, T\}$ denote the set of DNA alphabets. A sequence $Seq = ATGTCATG$ can be formed by arranging these symbols from Σ.

This notation is foundational in bioinformatics, where symbolic sequences like DNA, RNA, or protein sequences are studied to uncover meaningful patterns and biological insights.

13.3 Practical Representation

A symbolic sequence is typically stored as a file on a computer. To ensure consistency and readability, follow these guidelines when creating and managing the file:

- **Enter Sequence in Line**: Write the entire sequence of symbols on a single line without breaks.
- **No Delimiter Between the Symbols**: Do not insert any delimiters between consecutive symbols in the sequence.

Thus, the format of a sequence in this representation is

$$symbol_1 symbol_2 symbol_3 \cdots symbol_n$$

Example 13.2 An example of a symbolic sequence representing a DNA sequence $S = ACTGCATGCTATGCATGC$.

13.4 Creating Synthetic Symbolic Sequence Databases

The PAMI package provides a robust and adaptable tool for generating synthetic symbolic sequential databases, ideal for testing and developing algorithms in fields like data mining. Users can customize these databases to meet specific needs, such as defining the length of a sequence and the total number of symbols, making them invaluable resources for benchmarking and experimentation.

The following example code demonstrates how to create a synthetic symbolic sequential database using PAMI:

Program 1: Generating Synthetic Symbolic Sequential Database

```
from PAMI.extras.syntheticDataGenerator \
    import symbolicSequenceDatabase as db

obj = db.symbolicSequenceDatabase(
        sequenceSize=100000,
        numberOfSymbols=10
        )
obj.create()
obj.save('symbolicSequentialDB.csv')
#read the generated sequences into a dataframe
symbolicSequentialDataFrame=obj.getTransactions()
#stats
print('Runtime: ' + str(obj.getRuntime()))
print('Memory (RSS): ' + str(obj.getMemoryRSS()))
print('Memory (USS): ' + str(obj.getMemoryUSS()))
```

The above code snippet creates a symbolic sequence database of 100,000 symbols using 10 distinct symbols. The generated data is saved to a CSV file and loaded into a dataframe for further analysis. Additionally, runtime and memory statistics are printed, providing useful performance metrics for evaluation.

In specific real-world scenarios, users are interested in generating synthetic DNA and RNA sequences. PAMI library contains programs that facilitate the creation of synthetic DNA or RNA sequences for users.

Program 2: Generating Synthetic DNA/RNA Database

```
from PAMI.extras.syntheticDataGenerator \
    import NucleotideSequence as db
```

```
obj = db.NucleotideSequence(
        sequenceLength=100000,
        gcContent=0.5
        )
obj.create()
obj.save('symbolicSequentialDB.csv')
#read the generated sequences into a dataframe
symbolicSequentialDataFrame=obj.getTransactions()
#stats
print('Runtime: ' + str(obj.getRuntime()))
print('Memory (RSS): ' + str(obj.getMemoryRSS()))
print('Memory (USS): ' + str(obj.getMemoryUSS()))
```

13.5 Knowing the Statistical Details

The `dbStats` sub-sub-package within the `extras` package of PAMI allows users to obtain essential statistical details about a symbolic sequential database. This functionality is particularly useful for understanding database properties, which can be critical for various data analysis tasks. The `dbStats` package provides the following statistics:

1. **Total Number of Symbols**: Counts the distinct symbols within the database.
2. **Total Size of a Sequence**: Calculates the overall length of the sequence.
3. **Symbol Frequencies**: Determines how frequently each symbol has appeared.

Here is an example demonstrating how to use `dbStats` to retrieve these statistics:

Program 3: Deriving Statistical Details

```
from PAMI.extras.stats import \
    SymbolcSequentialDatabase as stat

# Create an instance of the statistical analysis class
obj = stat.SymbolcSequentialDatabase("symbolicSequentialDB.csv")

# Run the statistical analysis
obj.run()
```

```
# Print the statistical details
obj.printStats()

# Plot graphs for visual analysis
obj.plotGraphs()
```

In this example, the code loads a previously generated symbolic sequence database from a CSV file. It then runs a statistical analysis to gather and print the database's statistical details. Additionally, it visualizes the data through graphs, allowing for an insightful look at symbol distributions and sequence characteristics.

13.6 Frequent Contiguous Patterns

13.6.1 Basic Model

Definition 13.1 (Contiguous Pattern) A **contiguous pattern** $P \subseteq S$ is formally defined as $\langle s_j, s_{j+1}, \ldots, s_{j+k-1} \rangle$, where $1 \leq j \leq j + k - 1 \leq n$. This definition implies that the elements of P occupy consecutive positions in S without any gaps, preserving the order of occurrences of symbols in the original sequence. If $|P| = k$, then P is called a *k-length contiguous pattern*.

Example 13.3 Let $P = ATG$ be a contiguous pattern such that $P \subseteq S$. Since P contains three alphabets (or $|P| = 3$), it is called a 3-length contiguous pattern.

Definition 13.2 (Support of a Pattern) The number of distinct occurrences of P in S represents its *support* and is denoted as support(P).

Example 13.4 The pattern P appears in S at three locations whose index positions are (5,7), (10,12), and (14,16). Thus, the *support* of P in S, i.e., support$(P) = 3$.

Definition 13.3 (Frequent Contiguous Pattern) A contiguous pattern P is said to be a **frequent contiguous pattern** if support$(P) \geq$ minSup, where `minSup` is a hyper-parameter that represents the user-specified minimum support value.

Example 13.5 If the user-specified minSup $= 2$, P is a frequent contiguous pattern as support$(P) \geq$ minSup.

Definition 13.4 (Problem Definition) Given a symbolic sequence database S and the user-specified *minimum support* (`minSup`), the **problem definition** of frequent contiguous pattern mining is to find the complete set of frequent contiguous patterns in S that have *support* no less than the `minsup` value.

13.6.2 Mining Algorithm

Several algorithms have been developed to identify frequent contiguous patterns in symbolic sequence data. The PAMI library implements the `PositionMining` algorithm. The following code demonstrates using `PositionMining` to discover frequent contiguous patterns.

Program 4: Frequent Contiguous Pattern Discovery

```
from PAMI.contiguousFrequentPattern.basic \
    import PositionMining as alg

# Initialize PositionMining algorithm
obj = alg.PositionMining(iFile='symbolicSequentialDB.csv',\
        minSup=100, delimiter='\t')
obj.mine()

# Retrieve discovered patterns and resource usage
Patterns = obj.getPatterns()
print("#Frequent Sequence Patterns:",len(Patterns))

# Display memory and runtime statistics
memUSS = obj.getMemoryUSS()
print("Total Memory in USS:", memUSS)
memRSS = obj.getMemoryRSS()
print("Total Memory in RSS:", memRSS)
runTime = obj.getRuntime()
print("Total Execution Time in ms:", runTime)
```

In this example, the `PositionMining` algorithm is initialized with an input file containing symbolic sequence data (`symbolicSequentialDB.csv`), a minimum support threshold (`minSup=100`), and a specified delimiter. The algorithm then mines for frequent contiguous patterns, which are stored in `Patterns`. Memory usage and runtime statistics are also displayed, providing insights into the algorithm's resource requirements.

13.7 Conclusion

This chapter explored symbolic sequence databases, which are essential for bioinformatics, data mining, and pattern discovery applications. We covered the theoretical and practical representations of symbolic sequences, techniques for synthetic database generation, and methods for analyzing database statistics. Additionally, we introduced contiguous pattern discovery, a powerful approach for identifying frequently occurring patterns within sequence data, which can reveal significant insights in large datasets.

Chapter 14
Pattern Discovery in Fuzzy Databases

Abstract This chapter introduces fuzzy databases as an advanced method for handling and analyzing data with uncertainty, distinguishing them from traditional transactional databases. By utilizing fuzzy membership functions, fuzzy databases transform utility databases into representations where data is stored with degrees of certainty, enabling more flexible analysis. The chapter discusses theoretical representations of fuzzy databases, practical considerations for database implementation, and methods for identifying fuzzy frequent patterns.

14.1 Introduction

This chapter provides an in-depth exploration of fuzzy databases, a type of database that supports data representation with varying levels of uncertainty, unlike traditional transactional or temporal databases where data is precisely defined. Fuzzy databases are derived from utility databases using fuzzy membership functions to capture the nuanced and uncertain nature of certain data. The chapter covers the following three core topics:

1. **Theoretical Representation**: Defines fuzzy transactional databases with formal mathematical structures, including fuzzy membership functions, fuzzy terms, and set theory.
2. **Practical Representation**: Describes the storage and organization of fuzzy databases in computing systems, focusing on the practical aspects of data handling.
3. **Pattern Discovery**: Outlines techniques to identify and analyze frequent fuzzy patterns within these databases.

This chapter combines theoretical insights with practical approaches, providing readers with essential tools and methodologies for working with fuzzy databases and uncovering valuable patterns in uncertain data environments.

U. K. Rage, *Hands-on Pattern Mining*,
https://doi.org/10.1007/978-981-96-6791-8_14

14.2 Theoretical Representation

Let $I = \{i_1, i_2, \ldots, i_m\}$, where $m \geq 1$, be a finite set of m distinct items (or attributes). A utility database, UD, is an ordered collection of transactions paired with unique transaction identifiers. Each transaction in this database contains items and their corresponding quantities. Specifically, we define $UD = \{(1, T_1), (2, T_2), \ldots, (ts, T_{ts})\}$, where $ts \in \mathbb{R}^+$ represents a timestamp, and each transaction $T_q \in UD$, for $1 \leq q \leq ts$, is a subset of I that contains multiple items with associated purchase quantities v_{i_q} for each item $i_q \in T_q$.

Example 14.1 Let $I = \{a, b, c, d, e, f\}$ be the set of items (or sensors measuring the concentrations of an air pollutant, say PM2.5[1]). A hypothetical utility database generated from recording the items in I is shown in Table 14.1. This database contains 12 transactions. Each transaction in this database is associated with a transaction identifier (tid). In the first transaction, $(1, \{a : 60, b : 65, d : 55\})$, 1 represents the transactional identifier, and $\{a : 60, b : 65, d : 55\}$ represents the transaction containing items and their associated quantities. This means that the sensors a, b, and d have recorded the PM2.5 values of 60, 65, and 55, respectively. Other sensors have not recorded any value for PM2.5.

Definition 14.1 Let $\{1, 2, \cdots, h\}$ be the set of fuzzy terms for a membership function μ. The set of linguistic variables that can be drawn from the membership function μ for an item i, denoted as $R_i = \{R_{i1}, R_{i2}, \cdots, R_{ih}\}$, where R_{ik}, $1 \leq k \leq h$, is the fuzzy term mapped to an item i.

Example 14.2 The set of fuzzy terms for the utility database shown in Table 14.1 are G, M, UH4SG, UH, VUH, and H (see Fig. 14.1a). Consequently, the set of fuzzy terms for an item a in Table 14.1, i.e., $R_a = \{a.G,\ a.M,\ a.UH4SG,\ a.UH,\ a.VUH, a.H\}$. The same can be stated for the remaining items in the table.

Definition 14.2 Let v_{iq} denote the quantitative value of an item i in the transaction T_q. The fuzzy set, denoted as f_{iq}, is the set of fuzzy terms with their membership

Table 14.1 Running example: utility database

tid	$itemset$	tid	$itemset$
1	$a : 60, b : 65, d : 55$	7	$a : 45, b : 60, c : 45, e : 25$
2	$a : 30, b : 70, e : 60$	8	$a : 55, d : 60$
3	$a : 55, c : 20$	9	$a : 60, b : 65, d : 30$
4	$a : 60, b : 65, d : 55$	10	$a : 45, d : 40, f : 40$
5	$a : 55, d : 60, f : 30$	11	$a : 60, b : 55, c : 65, d : 55$
6	$b : 55, c : 40, e : 45$	12	$b : 45, e : 65$

[1] PM2.5 represents the particulate matter that has a diameter of 2.5 micrometers or smaller.

$PM_{2.5}$ (μg/m³)	Symbol	Health Descriptor
~10	G	Good
11~15	M	Moderate
16~35	UH4SG	Unhealth for selective groups
36~50	UH	Unhealthy
51~70	VUH	Very Unhealthy
71~	H	Hazardous

$PM_{2.5}$ categories

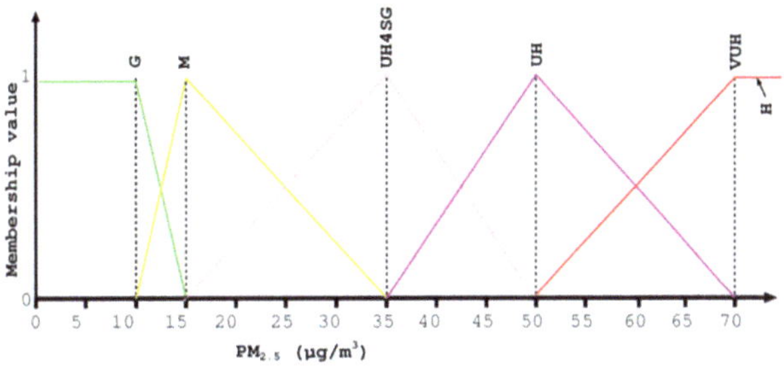

Fuzzy membership functions

Fig. 14.1 Graphical representation of fuzzy membership function for $PM_{2.5}$

degrees (fuzzy values) transformed from the quantitative value v_{iq} of the linguistic variable i by the membership functions μ as

$$\begin{aligned} f_{iq} &= \mu_i(v_{iq}) \\ &= \frac{fv_{iq1}}{R_{i1}} + \frac{fv_{iq2}}{R_{i2}} + \cdots + \frac{fv_{iqh}}{R_{ih}}, \end{aligned} \tag{14.1}$$

where h is the number of fuzzy terms of i transformed by μ, R_{il} is the lth fuzzy term of i, fv_{iql} is the membership degree (fuzzy value) of v_{iq} of i in the lth fuzzy term R_{il}, and $fv_{iql} \in [0, 1]$.

Example 14.3 Consider the item a in Table 14.1. The quantity of a in the first transaction is 60. Thus, $v_{a1} = 60$. Based on the membership function shown in Fig. 14.1, the fuzzy set of a in T_1, i.e.,

$$f_{a1} = \frac{0}{a.G} + \frac{0}{a.M} + \frac{0}{a.UH4SG} + \frac{0.5}{a.UH} + \frac{0.5}{a.VUH} + \frac{0}{a.H} = \frac{0.5}{a.UH} + \frac{0.5}{a.VUH}.$$

For simplicity, we represent $f_{a1} = \{a.UH : 0.5, a.VUH : 0.5\}$. The fuzzy transactional database [1, 2] generated from Table 14.1 is shown in Table 14.2.

14.3 Practical Representation

A fuzzy transactional database is typically stored as a file on a computer. To ensure consistency and readability, follow these guidelines when creating and managing the file:

- **Enter Transaction in Line**: Each transaction is written as a line.
- **Appearance of Fuzzy Terms**: Fuzzy terms appear at the beginning of the line. A delimiter, say tab space, separates the fuzzy terms.

Table 14.2 Fuzzy temporal database generated from Table 14.1

tid	$itemset$
1	$a.UH : 0.5, a.VUH : 0.5, b.UH : 0.25, b.VUH : 0.75, d.UH : 0.75, d.VUH : 0.25$
2	$a.M : 0.25, a.UH4G : 0.75, b.VUH : 1, e.UH : 0.5, e.VUH : 0.5$
3	$a.UH : 0.75, a.VUH : 0.25, c.M : 0.75, c.UH4G : 0.25$
4	$a.UH : 0.5, a.VUH : 0.5, b.UH : 0.25, b.VUH : 0.75, d.UH : 0.75, d.VUH : 0.25$
5	$a.UH : 0.75, a.VUH : 0.25, d.UH : 0.5, d.VUH : 0.5, f.M : 0.5, f.UH4G : 0.5$
6	$b.UH : 0.75, b.VUH : 0.25, c.UH4G : 0.66, c.UH : 0.33, e.UH4G : 0.33, e.UH : 0.66$
7	$a.UH4G : 0.33, a.UH : 0.66, b.UH : 0.5, b.VUH : 0.5, c.UH4G : 0.33, c.UH : 66,$ $e.M : 0.5, e.UH4G : 0.5$
8	$a.UH : 0.75, a.VUH : 0.25, d.UH : 0.5, d.VUH : 0.5$
9	$a.UH : 0.5, a.VUH : 0.5, b.UH : 0.25, b.VUH : 0.75, d.M :, d.UH4G :$
10	$a.UH4G : 0.33, a.UH : 0.66, d.UH4G : 0.66, d.UH : 0.33, f.UH4G : 0.66,$ $f.UH : 0.33$
11	$a.UH : 0.5, a.VUH : 0.5, b.UH : 0.75, b.VUH : 0.25, c.UH : 0.25, c.VUH : 0.75,$ $d.UH : 0.75, d.VUH : 0.25$
12	$b.UH4G : 0.33, b.UH : 0.66, e.UH4G : 0.33, e.UH : 0.66$

- **Appearance of Fuzzy Values**: Fuzzy values appear after the fuzzy terms. A delimiter, say tab space, separates the fuzzy value. The delimiter for fuzzy terms and fuzzy values must remain the same.
- **Delimiter for Fuzzy Terms and Fuzzy Values**: The fuzzy terms and fuzzy values in a transaction are separated with a colon mark (:) as a delimiter. This delimiter is fixed and cannot be overwritten by the users.
- **Neglecting TID Information**: Since each row represents the tid information of a transaction, we do have to store the tid information in the file.

Thus, the format of a sequence in this representation is

$$fuzzTerm_1 fuzzTerm_2 \cdots fuzzTerm_n : fuzzVal_1 fuzzVal_2 \cdots fuzzVal_n.$$

Example 14.4 An example of the first transaction appearing in the fuzzy transactional database shown in Table 14.2 is a.UH a.VUH b.UH b.VUH d.UH d.VUH:0.5 0.5 0.25 0.75 0.75 0.25.

14.4 Fuzzy Frequent Patterns

14.4.1 Basic Model

Definition 14.3 (The Support of a Fuzzy Term) Let FTD' denote the fuzzy transactional database generated from the UD using the fuzzy membership function

μ. The *support* of the transformed fuzzy terms, denoted $sup(R_{il})$, is the summation of scalar cardinality of the fuzzy values of fuzzy term R_{il}, which can be defined as

$$sup(R_{il}) = \sum_{R_{il} \subseteq T_q \wedge T_q \in FTD'} fv_{ilq}. \tag{14.2}$$

Example 14.5 Table 14.2 shows the fuzzy transactional database generated for the utility database shown in Table 14.1. The item $d.UH$ appears in the transactions whose timestamps are 1, 4, 5, 8, 10, and 11. Thus, the *support* of item $d.UH$, i.e., sup($R_{d.UH}$)=$fv_{d.UH_1} + fv_{d.UH_4} + fv_{d.UH_5} + fv_{d.UH_8} + fv_{d.UH_{10}} + fv_{d.UH_{11}}$=0.75 + 0.75 + 0.5 + 0.5 + 0.33 + 0.75=3.58.

Definition 14.4 (The Support of a Fuzzy k-Pattern) The support of fuzzy k-pattern ($k \geq 2$), denoted as $sup(X)$, is the summation of scalar cardinality of the fuzzy values for X, which can be defined as

$$sup(X) = \{X \in R_{il} | \sum_{R_{il} \subseteq T_q \wedge T_q \in FTD'} min(fv_{aql}, fv_{bql}), \tag{14.3}$$

where $a, b \in X$ and $a \neq b$.

Example 14.6 The set of fuzzy terms, $\{a.UH, d.UH\}$, is an itemset (or a pattern). This pattern contains two items. Therefore, it is a 2-pattern. In Table 14.2, the pattern $\{a.UH, d.UH\}$ occurs in the transactions whose transactional identifiers are 1, 4, 5, 8, 10, and 11. Thus, the *support* of $\{a.UH, d.UH\}$ in Table 14.2, i.e., $sup(a.UH, d.UH)$=$min(fv_{a.UH_1}, fv_{d.UH_1}) + min(fv_{a.UH_4}, fv_{d.UH_4}) + min(fv_{a.UH_5}, fv_{d.UH_5}) + min(fv_{a.UH_8}, fv_{d.UH_8}) + min(fv_{a.UH_{10}}, fv_{d.UH_{10}}) + min(fv_{a.UH_{11}}, fv_{d.UH_{11}}) = min(0.5, 0.75) + min(0.75, 0.5) + min(0.75, 0.5) + min(0.75, 0.5) + min(0.66, 0.33) + min(0.5, 0.75) = 0.5 + 0.5 + 0.5 + 0.5 + 0.33 + 0.5 = 2.83$.

Definition 14.5 (Fuzzy Frequent Pattern X) A pattern X is called a fuzzy frequent pattern if its *support* is no less than the user-specified *minimum support* ($minSup$). In other words, X is a fuzzy frequent pattern if $sup(X) \geq minSup$.

Example 14.7 If the user-specified $minSup = 2$, then the fuzzy pattern $\{a.UH, d.UH\}$ is said to be a fuzzy frequent pattern because $sup(\{a.UH, d.UH\}) \geq minSup$. The above pattern provides useful information that the sensors a and d have frequently observed unhealthy levels of $PM_{2.5}$.

Definition 14.6 (Problem Definition) Given the quantitative transactional (or utility) database (UD), the user-specified fuzzy membership functions (μ), and *minimum support* ($minSup$) value, the problem of fuzzy frequent pattern mining involves discovering all patterns in FTD' that have $sup(X) \geq minSup$.

14.4.2 Mining Algorithm

Several algorithms have been developed to identify fuzzy frequent patterns in fuzzy transactional database. The PAMI library implements the `FFI-Miner` [2] algorithm. The following code demonstrates using `FFI-Miner` to discover fuzzy frequent patterns.

Program 1: Fuzzy Frequent Patterns in Fuzzy Transactional Database

```
from PAMI.fuzzyFrequentPattern.basic \
    import FFIMiner  as alg

inputFile = 'Fuzzy_T10I4D100K.csv'
minimumSupportCount = 100

obj = alg.FFIMiner(iFile=inputFile, \
    minSup=minimumSupportCount, sep='\t')
obj.mine()

# Retrieve discovered patterns and resource usage
obj.save('fuzzyfrequentPatterns.txt')
Patterns = obj.getPatterns()
print("Total number of Fuzzy Frequent Patterns:",len(Patterns))

# Display memory and runtime statistics
memUSS = obj.getMemoryUSS()
print("Total Memory in USS:", memUSS)
memRSS = obj.getMemoryRSS()
print("Total Memory in RSS:", memRSS)
runTime = obj.getRuntime()
print("Total Execution Time in ms:", runTime)
```

In this example, the `FFI-Miner` algorithm is initialized with an input file containing fuzzy transactional database (`Fuzzy_T10I4D100K.csv`), a minimum support threshold (`minSup=100`), and a specified delimiter. The algorithm then mines for fuzzy frequent patterns, which are stored in `Patterns`. Memory usage and runtime statistics are also displayed, providing insights into the algorithm's resource requirements.

14.5 Other Types of Fuzzy Databases

This chapter primarily covered fuzzy transactional databases and how to mine them. However, other forms of fuzzy databases exist in the real world. For example, adding the timestamp information to a fuzzy transactional database will result in a fuzzy temporal database. Similarly, considering the spatial information of the items in the fuzzy transactional (or temporal) database will result in a fuzzy spatio-transactional (or spatiotemporal) database. Users can find interesting patterns, such as fuzzy periodic-frequent patterns, fuzzy geo-referenced frequent patterns, and fuzzy geo-referenced periodic-frequent patterns from these databases.

14.6 Conclusion

This chapter presented fuzzy databases as a method for managing and analyzing data with inherent uncertainty, offering an alternative to traditional transactional databases. Using fuzzy membership functions, we can transform utility databases into fuzzy transactional databases, allowing data to be represented in degrees rather than absolute values. This chapter covered the theoretical and practical frameworks for fuzzy databases, including the representation of fuzzy data and techniques for mining fuzzy frequent patterns. The implementation code accompanying the examples in this chapter can be accessed on our GitHub repository: https://github.com/UdayLab/Hands-on-Pattern-Mining/blob/main/chapter14.ipynb.

References

1. C. Lin, T. Hong and W. Lu. Linguistic data mining with fuzzy FP-trees. In *Expert Systems with Applications*, pp. 4560–4567, 2010.
2. C.-W. Lin, T. Li, P. Fournier Viger and T.-P. Hong. A fast algorithm for mining fuzzy frequent itemsets. Journal of Intelligent and Fuzzy Systems, vol. 29, no. 6, pp. 2373–2379, 2015.

Chapter 15
Knowledge Discovery in Uncertain Databases

Abstract This chapter focuses on uncertain transactional databases (UTDB), which represent collections of transactions containing items with associated probabilities, reflecting the inherent uncertainty in real-world data. We formally define uncertain transactions, patterns, and their expected support in the context of uncertain data. The chapter also covers techniques for creating synthetic uncertain transactional databases, converting structured data into UTDBs, and deriving statistical details to understand the data's characteristics better. A significant portion of the chapter is dedicated to the challenges of frequent pattern discovery, specifically addressing the limitations of the downward closure property in uncertain data and introducing algorithms like TUBE-P for efficient pattern mining. Practical Python code examples demonstrate how these methods can be implemented to analyze uncertain transactional data.

15.1 Introduction

An uncertain transactional database is a collection of unordered transactions, where each transaction consists of items along with their associated occurrence probabilities. This type of data is commonly found in real-world scenarios such as sales, healthcare, clickstream analysis, and sensor networks, where uncertainty about the presence of items in a transaction is inherent. The diagram in Fig. 15.1 illustrates how various factors combine to form an uncertain transactional database, emphasizing the complex relationships involved.

Other types of uncertain transactional databases exist, such as uncertain temporal databases, uncertain geo-referenced transactional databases, and uncertain utility databases, each incorporating different factors. This chapter focuses on introducing uncertain transactional databases and exploring methods for discovering interesting patterns within them, considering the inherent uncertainty in the data.

This chapter addresses the following key aspects of uncertain transactional databases:

U. K. Rage, *Hands-on Pattern Mining*,
https://doi.org/10.1007/978-981-96-6791-8_15

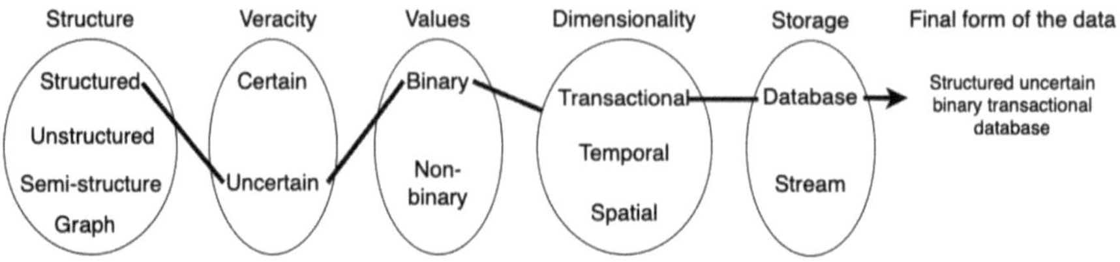

Fig. 15.1 Illustration of factors contributing to the creation of an uncertain transactional database

1. **Theoretical Representation**: The formal definition of an uncertain transactional database using set theory.
2. **Practical Representation**: Implementing and storing uncertain transactional databases within computer systems.
3. **Synthetic Database Creation**: Methods for generating synthetic uncertain transactional databases used for testing and benchmarking.
4. **Dataframe Conversion**: Techniques for transforming structured dataframes into uncertain transactional databases, enabling broader data analysis applications.
5. **Database Statistics**: Approaches for deriving statistical insights from an uncertain transactional database.
6. **Finding Frequent Patterns**: A formal definition of frequent patterns and a detailed procedure for discovering them in uncertain transactional databases.

15.2 Theoretical Representation

Let $I = \{i_1, i_2, \ldots, i_n\}$, where $n \geq 1$, be a set of items. Let $X \subseteq I$ represent an itemset (or a pattern). A pattern that contains k items is called a k-pattern.

An uncertain transaction, denoted t_{tid}, consists of a transaction identifier (tid) and a pattern Y. That is, $t_{tid} = (tid, Y)$, where $Y \subseteq I$ is the set of items in the pattern. Importantly, each item $i_k \in Y$ is associated with an existential probability $P(i_k, t_{tid}) \in (0, 1)$, which represents the likelihood of the presence of item i_k in the uncertain transaction t_{tid}.

An uncertain transactional database [1], denoted $UTDB$, is a collection of such uncertain transactions:

$$UTDB = \{t_1, t_2, \ldots, t_m\}, \quad m \geq 1.$$

Each transaction in the database is associated with a transaction identifier, the corresponding pattern, and the probabilities for each item in the pattern. This structure allows for the representation of uncertainty regarding the presence of items in each transaction.

Table 15.1 Uncertain transactional database

tid	Transaction
1	b(0.1) c(0.8) d(0.9)
2	a(0.7) c(0.7) d(0.1)
3	a(0.8) b(0.6) c(0.4)
4	c(0.3) d(0.4) e(0.9)

Example 15.1 Let $I = \{a, b, c, d, e\}$ be a set of fixed sensors (or items). A hypothetical uncertain transactional database constituting these items is shown in Table 15.1. The set of items a and c, i.e., $\{a, c\}$ (or ac, in short) is a pattern. This is a 2-pattern as it contains only two items.

15.3 Practical Representation

An uncertain transactional database is usually stored as a file on a computer. To properly create and manage this file, follow these four rules:

- **One Transaction per Line**: Each line in the file represents a single transaction. The line number implicitly acts as the transaction identifier (tid), so it is not explicitly stored in the file to save space and reduce processing costs.
- **Order of Occurrences**: All items occur first in a transaction. Next, the uncertainty values appear in the same order as the items have occurred. Items and their uncertainty values are separated with a fixed delimiter, a colon mark.
- **Unique Items per Transaction**: Each item should appear once per line. The items can be listed in any order within the line.
- **A Delimiter Separates Items and Uncertain Values**: Items and their uncertain values in a transaction are separated by a delimiter, such as a space or tab. The PAMI algorithms use a `tab` as the default delimiter, but users can choose other delimiters like commas or spaces.

Overall, the format of a transaction in an uncertain transactional database is

$$item_1\langle sep\rangle item_2\langle sep\rangle \cdots : value_1\langle sep\rangle value_2\langle sep\rangle \cdots$$

Example 15.2 If the delimiters are a `tab` and a colon mark, the uncertain transactional database shown in Table 15.1 would look like this:

```
b	c	d:0.1	0.8	0.9
a	c	d:0.7	0.7	0.1
a	b	c:0.8	0.6	0.4
c	d	e:0.3	0.4	0.9
```

15.4 Creating Synthetic Uncertain Transactional Database

The PAMI package offers a powerful and flexible tool for generating uncertain synthetic transactional databases tailored to various needs. This capability is invaluable for testing and developing algorithms in data mining and related fields. Users can customize the database to suit their specific requirements, including the number of transactions, the total number of items, and the average transaction length.

To illustrate the creation of an uncertain synthetic transactional database, consider the following sample code. This example generates a database with 100,000 transactions, each containing an average of 10 items from a set of 1,000 possible items:

Program 1: Generating Synthetic Uncertain Transactional Database

```
from PAMI.extras.syntheticDataGenerator \
    import UncertainTransactionalDatabase as db

obj = db.UncertainTransactionalDatabase(
        databaseSize=100000,
        avgItemsPerTransaction=10,
        numItems=1000,
        sep='\t'
        )
obj.create()
obj.save('uncertainTDB.csv')
#read the generated transactions into a dataframe
transactionalDataFrame=obj.getTransactions()
#stats
print('Runtime: ' + str(obj.getRuntime()))
print('Memory (RSS): ' + str(obj.getMemoryRSS()))
print('Memory (USS): ' + str(obj.getMemoryUSS()))
```

15.5 Converting a Dataframe into an Uncertain Transactional Database

The PAMI package provides a convenient method to convert a dataframe into an uncertain transactional database, particularly useful for transaction-based data analysis. The following Python code demonstrates how to perform this conversion:

Program 2: Dataframe to Uncertain Transactional Database Conversion

```
from PAMI.extras.convert import DF2DB as alg
import pandas as pd
import numpy as np

# Creating a 100 x 4 DataFrame with random values
data = np.random.uniform(0, 1, size=(100, 4))
dataFrame = pd.DataFrame(data,
                columns=['Item1', 'Item2', 'Item3', 'Item4']
               )

# Converting the DataFrame to an uncertain transactional database
# by considering values greater than or equal to a threshold (0.6)
obj = alg.DF2DB(dataFrame)
obj.convert2UncertainTransactionalDatabase(
        oFile='UTDB.csv',
        condition='>=', thresholdValue=0.6
     )

# Printing runtime and memory usage statistics
print('Runtime: ' + str(obj.getRuntime()))
print('Memory (RSS): ' + str(obj.getMemoryRSS()))
print('Memory (USS): ' + str(obj.getMemoryUSS()))
```

15.6 Obtaining Statistical Details

The `dbStats` sub-package in PAMI's `extras` module allows users to retrieve statistical details about an uncertain transactional database. These statistics are important for understanding the underlying properties of the database, which can inform various data analysis tasks. The statistical details provided by `dbStats` include:

1. Database size
2. Total number of items in the database
3. Minimum, average, and maximum lengths of the transactions
4. Standard deviation of transaction sizes
5. Variance in transaction sizes
6. Sparsity of the database

7. Frequencies of the items in the database
8. Distribution of transaction lengths

Below is an example demonstrating how to use dbStats to derive these statistics from an uncertain transactional database:

Program 3: Deriving the Statistical Details

```
from PAMI.extras.dbStats import \
    UncertainTransactionalDatabase as stat

# Load the uncertain transactional database
obj = stat.UncertainTransactionalDatabase("UTDB.csv")

# Run the statistics generation
obj.run()

# Print the calculated statistics
obj.printStats()

# Plot graphical representations of the statistics
obj.plotGraphs()
```

15.7 Frequent Pattern Discovery

15.7.1 Basic Model

Definition 15.1 (Expected Support of Pattern X in a Transaction) The existential probability of X in t_{tid}, denoted as $P(X, t_{tid})$, represents the product of corresponding existential probability values of all items in X when these items are independent. That is,

$$P(X, t_{tid}) = \prod_{\forall i_j \in X} P(i_j, t_{tid}).$$

The expected support of X in the uncertain transactional database $UTDB$, denoted as $expSup(X)$, is given by

$$expSup(X) = \sum_{tid=1}^{m} P(X, t_{tid}),$$

where m is the total number of transactions in the database.

Example 15.3 Consider the pattern ac, which occurs in transactions with $tids$ of 2 and 3. The existential probability of ac in the second transaction is

$$P(ac, t_2) = P(a, t_2) \times P(c, t_2) = 0.7 \times 0.7 = 0.49.$$

Similarly, the existential probability of ac in the third transaction is

$$P(ac, t_3) = 0.32.$$

The expected support of ac in the entire database is

$$expSup(ac) = 0.49 + 0.32 = 0.81.$$

Definition 15.2 (Frequent Pattern X**)** A pattern X is considered frequent if its expected support satisfies the condition:

$$expSup(X) \geq minSup,$$

where $minSup$ represents the user-specified minimum support threshold.

Example 15.4 Suppose the user specifies a minimum support value of $minSup = 0.6$. In that case, the pattern ac is considered frequent since its expected support, $expSup(ac) = 0.81$, is greater than or equal to the minimum support threshold.

15.7.2 Search Space

The set of items in a database forms an itemset lattice. This lattice represents the search space for pattern discovery in certain and uncertain transactional data. The search space size is $2^n - 1$, where n represents the total number of items in the database.

15.7.3 Inability of Apriori Property

Although the search space for frequent pattern discovery is the same for certain and uncertain data, the computational cost for finding these patterns differs. The reason is as follows:

The frequent patterns discovered from certain data satisfy the downward closure property. This property plays a key role in reducing the computational costs of finding frequent patterns in certain data. However, the frequent patterns discovered from uncertain data do not satisfy the downward closure property. This increases the search space, leading to a higher computational cost in finding frequent patterns in uncertain data. To address this challenge, mining algorithms employ various upper-bound constraints to help find frequent patterns in uncertain data.

15.7.4 Finding Frequent Patterns

Several algorithms, such as PUF [2], TUBE-P [3], and TUBE-S [3], have been proposed in the literature to find frequent patterns in uncertain transactional databases. While there is no universally accepted best algorithm, TUBE-P is widely used for its relatively faster performance than other algorithms. Below is a sample Python script demonstrating how to use the TUBE-P algorithm from the PAMI package to discover frequent patterns in an uncertain transactional database.

Program 1: Frequent Pattern Discovery Using TUBE-P

```
from PAMI.uncertainFrequentPattern.basic \
    import TubeP as alg

# Input file and minimum support count for frequent pattern mining
inputFile = 'uncertainTransaction_T10I4D100K.csv'
minSupport = 300

# Create an instance of the TubeP algorithm
obj = alg.TubeP(iFile=inputFile,
                minSup=minSupport, sep='\t')

# Mine frequent patterns
obj.mine()

# Save the discovered frequent patterns to a file
obj.save('frequentPatterns.txt')

# Convert the frequent patterns into a DataFrame
frequentPatternsDF = obj.getPatternsAsDataFrame()

# Display the number of frequent patterns and resource usage
print('#Patterns: ' + str(len(frequentPatternsDF)))
print('Runtime: ' + str(obj.getRuntime()))
```

```
print('Memory (RSS): ' + str(obj.getMemoryRSS()))
print('Memory (USS): ' + str(obj.getMemoryUSS()))
```

15.8 Conclusion

In this chapter, we explored the concept of uncertain transactional databases, highlighting their structure, the challenges in handling uncertainty, and the methods to mine frequent patterns from such data. We introduced the formalization of uncertain transactions and patterns, the process of generating synthetic uncertain data, and the conversion of structured dataframes into transactional databases. We also discussed the key statistical measures necessary for understanding the properties of uncertain transactional data. Given the complexity of uncertain data, we outlined the limitations posed by the inability to apply the downward closure property, which increases computational costs. We highlighted algorithms like TUBE-P that utilize upper-bound constraints to discover frequent patterns effectively. We demonstrated how to apply these techniques to real-world scenarios by providing practical Python code examples. In conclusion, while uncertain transactional data presents unique challenges, the methods and tools discussed offer valuable solutions for efficient analysis and pattern discovery, with future potential for optimization and application across various domains.

References

1. Chui, C. K., Kao, B., and Hung, E. (2007). Mining frequent itemsets from uncertain data. In: PAKDD, pp. 47–58 (2007).
2. Leung, C.K.S., Tanbeer, S.K.: PUF-Tree: a compact tree structure for frequent pattern mining of uncertain data. In: PAKDD, pp. 13–25 (2013).
3. Leung, C.K., MacKinnon, R.K., Tanbeer, S.K.: Fast algorithms for frequent itemset mining from uncertain data. In: ICDM, pp. 893–898 (2014).

Chapter 16
Finding Useful Patterns in Graph Databases

Abstract Graph transactional databases are essential for modeling complex relationships in various real-world applications, such as social networks, bioinformatics, and cheminformatics. These databases can be categorized as either certain or uncertain, depending on whether their edge connections are deterministic or probabilistic. This chapter provides a comprehensive exploration of graph transactional databases, covering both theoretical and practical representations. It introduces formal definitions of graph structures using set theory, details various data storage formats (traditional and compressed), and explains procedures for converting between these formats. Additionally, it presents methodologies for generating synthetic graph databases and deriving statistical insights. Furthermore, the chapter discusses frequent subgraph pattern discovery, a crucial task for uncovering recurring structures within graph data. The use of the PAMI package is highlighted throughout, offering practical implementations for database creation, visualization, and analysis.

16.1 Introduction

Graph data structures are versatile tools for representing relationships in real-world data applications. They are instrumental in scenarios where entities and their interactions are at the forefront, such as in social networks, chemical compounds, or protein interactions.

A graph database is a collection of graphs, enabling structured analysis and querying of the represented relationships. Graph databases are typically categorized into two types based on the certainty of their edge connections:

- **Graph Certain Binary Transactional Database (or Graph Transactional Database) [1]:** In this type of database, the relationships (edges) between nodes are deterministic. Specifically, the probability of an edge existing between any two nodes is either 0 or 1. This deterministic nature ensures that all connections are either confirmed (present) or absent, with no ambiguity.
- **Graph Uncertain Binary Transactional Database (or Uncertain Graph Transactional Database) [2]:** Unlike their certain counterparts, uncertain graph

U. K. Rage, *Hands-on Pattern Mining*,
https://doi.org/10.1007/978-981-96-6791-8_16

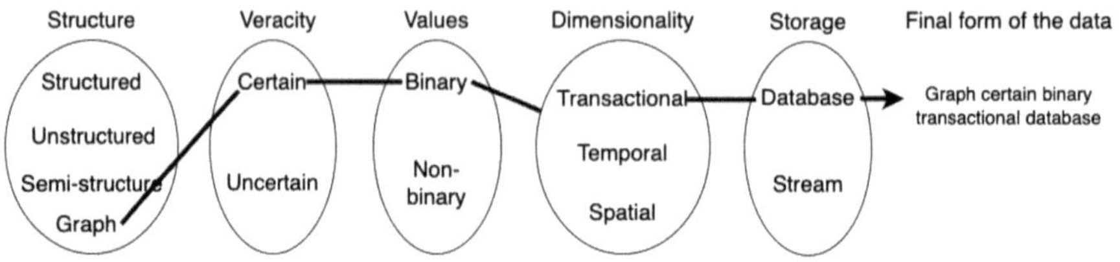

Fig. 16.1 Factors contributing to the creation of certain graph transactional database

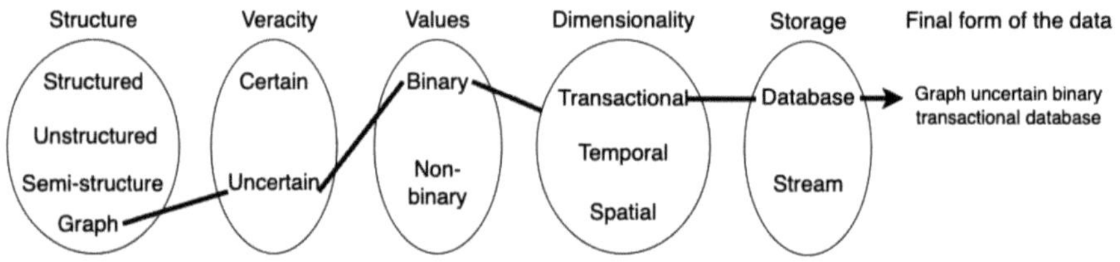

Fig. 16.2 Factors contributing to the creation of uncertain graph transactional database

transactional databases accommodate uncertainty in relationships. Here, the probability of an edge existing between two nodes lies in the range (0, 1). This probabilistic approach reflects situations where connections are not definitive but are associated with some degree of likelihood.

Figures 16.1 and 16.2 visually highlight how various factors influence the formation of these graph databases, illustrating the distinct characteristics of certainty and uncertainty within these data structures.

This chapter delves into graph transactional databases, focusing on their properties, representations, and methodologies for uncovering interesting patterns within them. The key topics covered in this chapter include:

1. **Theoretical Representation**: A rigorous formalization of graph databases using set theory, providing a foundation for understanding their structure and components.
2. **Practical Representation**: Details on how graph databases are stored and implemented in computer systems, emphasizing practical aspects of data representation and retrieval.
3. **Synthetic Graph Database Creation**: Techniques for generating synthetic graph databases, essential for testing algorithms, benchmarking methods, and evaluating performance in controlled environments.
4. **Graph Statistics**: How to derive statistical details about a graph database.
5. **Finding Frequent Subgraph Patterns**: Formal definitions and algorithms for identifying frequent subgraphs, which are graph patterns appearing recurrently across multiple graphs in the database.
6. **Finding Top-k Subgraph Patterns**: A formal framework and procedures for discovering the top-k subgraph patterns, focusing on ranking and retrieving the most significant patterns based on predefined criteria.

16.2 Theoretical Representation

Definition 16.1 (A Graph) An *exact graph* is formally defined as

$$G = (V, E, L_v, L_e, F_v, F_e),$$

where:

- V is the set of vertices (nodes).
- E is the set of edges (connections between nodes).
- L_v is the set of labels for vertices.
- L_e is the set of labels for edges.
- $F_v : V \rightarrow L_v$ maps each vertex to its corresponding label.
- $F_e : E \rightarrow L_e$ maps each edge to its corresponding label.

Example 16.1 Let:

- $L_v = \{A, B, C, D\}$: the set of vertex labels
- $L_e = \{1, 2, 3\}$: the set of edge labels
- $V = \{0, 1, 2, 3\}$: the identifiers of four vertices
- $E = \{\text{edge}(0, 1), \text{edge}(0, 2), \text{edge}(2, 3), \text{edge}(3, 1)\}$: the set of edges connecting pairs of vertices

The `Graph 1` in Fig. 16.3 represents an exact graph generated using L_v, L_e, V, and E.

Definition 16.2 (Graph Transactional Database) A *graph transactional database* is a collection of labeled exact graphs. It is denoted as

$$GTD = \{G_1, G_2, \ldots, G_n\},$$

where n is the number of graphs in the database, and each G_i is an exact graph.

Example 16.2 The collection of all four graphs illustrated in Fig. 16.3 constitutes a *graph transactional database*.

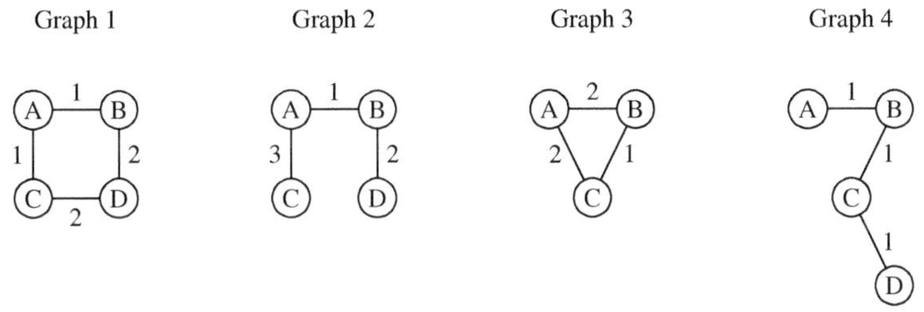

Fig. 16.3 An example of graph transactional database

16.3 Practical Representation

A graph transactional database is typically stored as a file on a computer. This database can be stored in two primary formats: the *traditional format* and the *compressed format*.

The **traditional format** is the most commonly used representation for storing graph databases. In this format, every graph's identifier, nodes, and edges are written on separate lines. However, the main limitation of this format is data redundancy, which can increase the memory and runtime requirements of mining algorithms.

The **compressed format**, introduced by PAMI researchers, represents each graph in a single line, listing nodes first, followed by edges and their labels. Algorithms in PAMI are designed to accept graph databases in both formats.

This subsection describes both formats and explains the procedures for converting between traditional and compressed formats.

16.3.1 Traditional Format

To prepare a graph transactional database in the traditional format, follow these rules:

- **Beginning a graph**: Start each graph with a triplet "`t # ` *id*" on a line, where $id \in (1, n)$ is an integer representing the graph identifier.
- **Storing a vertex**: A vertex in a graph is written as a triplet "`v vertexID vertexLabel`" on a line. The `vertexID` must be unique within each graph. The `vertexLabel` can appear multiple times in a graph, as multiple vertices can have the same label.
- **Storing the edges**: An edge in a graph is written as "`e vertexID_from vertexID_to edgeLabel`" on a line. The `vertexID_from` and `vertexID_to` indicate the vertices connected by the edge.

The overall structure of the traditional format is as follows:

```
t # graphID
v vertexID_1 vertexLabel
...
e vertexID_1 vertexID_2 edgeLabel
...
```

Example 16.3 The Graphs 1 and 2 from Fig. 16.3 can be represented in the traditional format as follows:

```
t # 0
v 0 A
v 1 B
v 2 C
```

```
v 3 D
e 0 1 1
e 1 3 2
e 3 2 2
e 2 0 1

t # 1
v 0 A
v 1 B
v 2 C
v 3 D
e 0 1 1
e 0 2 3
e 1 3 2
```

16.3.2 Compressed Format

To prepare a graph transactional database in a compressed format, follow these rules:

- **Each line represents a graph**: The information in a graph is stored in a single line.
- **Order of storing vertices and edges**: The vertices first appear in a line, followed by the edges. A fixed delimiter, a colon mark, separates the vertices and edges.
- **Vertex pairs**: In a line, each vertex is represented as a pair containing *vertex identifier* (vID) and *vertex label* (vL).
- **Edge triplets**: In a line, each edge is represented as a triplet containing *vertex identifier*, *vertex identifier*, and *edge label* (eL).

Overall, the compressed format of representing a graph transactional database is

```
vID_1 vL_1 vID_2 vL_2...:vID_1 vID_2 eLB_1 vID_1 vID_4 eL_2...
```

Example 16.4 The Graphs 1 and 2 in Fig. 16.3 are written as follows:

```
0 A 1 B 2 C 3 D:0 1 1 1 3 2 3 2 2 2 0 1
0 A 1 B 2 C 3 D:0 1 1 0 2 3 1 3 2
```

16.3.3 Procedures for Converting Traditional into Compressed Format

The PAMI library provides functionality to transform a graph transactional database from a traditional format into a compressed format and vice versa.

Program 1: Converting the Format

```
from PAMI.extras.graph \
    import GraphConvertor as gc

obj = gc.GraphConvertor(iFile='Chemical_340.txt')
obj.convertTraditional2Compressed()
#obj.convertFromCompressed2Traditional()
obj.save('compressedGraphData.csv')

# Stats
print('Runtime: ' + str(obj.getRuntime()))
print('Memory (RSS): ' + str(obj.getMemoryRSS()))
print('Memory (USS): ' + str(obj.getMemoryUSS()))
```

16.4 Creating Synthetic Graph Transactional Database

The PAMI package provides a flexible and efficient tool for generating graph transactional databases, which can be customized to meet specific needs. This feature is handy for testing and developing algorithms in data mining and related fields.

To demonstrate how to create a synthetic graph transactional database, the following sample code can be used:

Program 2: Generating Synthetic Graph Transactional Database

```
from PAMI.extras.syntheticDataGenerator import
↪   certainGraphTransactions as db

obj = db.certainGraphTransactions(numGraphs=100,
↪   avgNumVertices=10, avgNumEdges=6, numVertexLabels=5,
↪   numEdgeLabels=3, outputFileName='opn.txt', format='old')
obj.generate()
#stats
print('Runtime: ' + str(obj.getRuntime()))
print('Memory (RSS): ' + str(obj.getMemoryRSS()))
print('Memory (USS): ' + str(obj.getMemoryUSS()))
```

16.5 Visualizing the Graph Database

The PAMI package allows users to view graphs in a graph database. Below is a sample Python code for this purpose.

Program 3: Visualizing the Graph Database

```
from PAMI.extras.visualize import graphs as vis
objVis = vis.graphDatabase(iFile='graphTransactionalDB.csv')
objVis.plot()
```

16.6 Obtaining Statistical Details

The `stats` sub-package in PAMI's `extras` module allows users to retrieve statistical details about a graph database. These statistics are important for understanding the underlying properties of the database, which can inform various data analysis tasks. The statistical details provided by `stats` include:

1. Average number of nodes
2. Average number of edges
3. Minimum, average, and maximum number of nodes
4. Minimum, average, and maximum number of edges
5. Total number of unique vertex labels
6. Total number of unique edge labels

Below is an example demonstrating how to use `stats` to derive these statistics from an uncertain transactional database:

Program 4: Deriving the Statistical Details

```
from PAMI.extras.stats import graphDatabase as alg

# Load the uncertain transactional database
obj = alg.graphDatabase(iFile='Chemical_340.txt')

# Print the calculated statistics
obj.printGraphDatabaseStatistics()
obj.printIndividualGraphStats()

```

```
# Plot graphical representations of the statistics
obj.plotEdgeDistribution()
obj.plotNodeDistribution()
```

16.7 Frequent Subgraph Pattern Discovery

16.7.1 Basic Model

Definition 16.3 (A Subgraph) A subgraph, S in an exact graph, $G = (V, E, L_v, L_e, F_v, F_e)$ is defined as $S = (V_s, E_s, L_{sv}, L_{se}, F_{sv}, F_{se})$, such that $S \sqsubseteq G$, iff $V_s \subseteq V$ and $E_s \subseteq E$. A subgraph is a part of the graph.

Example 16.5 An example of a subgraph, S, is

```
t # 0
v 0 A
v 1 B
e 0 1 1
```

Definition 16.4 (Support of Subgraph Pattern) Support, sup, of a subgraph pattern S in a graph transactional dataset D, is defined as

$$sup(S) = \frac{|\{g \mid g \in D \wedge S \sqsubseteq g\}|}{|GTD|}.$$

It means that the subgraph, S, is isomorphic to some pattern that is a subset of graph g, for all such graphs which belong to GTD. It is finally normalized by dividing by GTD and keeping its range between 0 and 1.

Example 16.6 Continuing the above example, the subgraph S appears in three graphs (i.e., Graph 1 (G_1), Graph 2 (G_2), and Graph 4 (G_4)) of the graph transactional database shown in Fig. 16.3. Thus, the support of S, i.e., $sup(S) = |\{G_1, G_2, G_4\}|/|GTD| = 3/4 = 0.75$.

Definition 16.5 (Frequent Subgraph Pattern X) A subgraph pattern S is considered frequent if its support is no less than the user-specified *minimum support* ($minSup$) threshold value.

Example 16.7 If the user-specified $minSup = 0.5$, then S is a frequent subgraph as $sup(S) \geq minSup$.

16.7.2 Finding Frequent Subgraph Patterns

The literature describes several algorithms for finding frequent subgraph patterns. While there is no universally accepted best algorithm, gSpan [1] is widely used due to its relatively faster performance than other algorithms. Below is a sample Python script demonstrating how to use the gSpan algorithm from the PAMI package to discover frequent subgraph patterns in a graph transactional database.

Program 5: Frequent Subgraph Pattern Discovery Using GSpan

```python
from PAMI.subgraphMining.basic import gspan as alg
obj = alg.GSpan('Chemical_340.txt', minSupport=100)
obj.mine()
frequentGraphs = obj.getFrequentSubgraphs()
memUSS = obj.getMemoryUSS()
print("Total Memory in USS:", memUSS)
memRSS = obj.getMemoryRSS()
print("Total Memory in RSS", memRSS)
run = obj.getRuntime()
print("Total ExecutionTime in seconds:", run)
obj.save('frequentSupgraphs.txt')
```

16.7.3 Visualization of the Frequent Subgraphs

Since a subgraph represents the portion of a graph, one can visualize the generated frequent subgraphs using the below-provided Python code.

Program 6: Visualizing of the Frequent Subgraphs

```python
from PAMI.extras.visualize import graphs as vis
objVis = vis.graphDatabase(iFile='frequentSupgraphs.txt')
objVis.plot()
```

16.8 Top-*k* Subgraphs

Specifying an appropriate $minSup$ value to find frequent subgraphs is a nontrival and challenging task in graph mining. When confronted with this problem in the real-world applications, researchers tried to tackle it by mining top-k frequently occurring subgraphs in a graph transactional database. The motivating reason is that specifying k value is much easier than specifying the right $minSup$ value.

16.8.1 Basic Model

Definition 16.6 (Top-*k* Subgraphs) Let $P = \{S_1, S_2, \cdots, S_z\}, z \geq 1$, be an ordered set of all subgraphs that can be generated from a graph transactional database (GTD) such that $sup(S_1) > sup(S_2) \geq \cdots sup(S_z)$. Let $Q = \{S_1, S_2, \cdots, S_k\} \subseteq P, 1 \leq k \leq z$, denote the set of k subgraphs that have highest $support$. That is $\forall S_x \in P$, $sup(S_x) \geq max(sup(S_y|\forall S_y \in P - Q)$, where $1 \leq x \leq k \leq y \leq z$.

16.8.2 Finding Top-k Subgraphs

The PAMI library implements the popular TKG [3] algorithm to find top-k subgraphs in a graph transactional database. Below is a sample Python script demonstrating how to use the TKG algorithm from the PAMI package to discover top-k subgraph patterns in a graph transactional database.

Program 7: Top-*k* Subgraphs Using TKG

```python
from PAMI.subgraphMining.topK import tkg as alg
obj = alg.TKG(iFile='Chemical_340.txt',k=100)
obj.mine()
frequentGraphs = obj.getKSubgraphs()
memUSS = obj.getMemoryUSS()
print("Total Memory in USS:", memUSS)
memRSS = obj.getMemoryRSS()
print("Total Memory in RSS", memRSS)
run = obj.getRuntime()
print("Total ExecutionTime in seconds:", run)
obj.save('frequentTopkSubgraphs.txt')
```

16.8.3 Visualization of the Top-k Subgraphs

Since a subgraph represents the portion of a graph, one can visualize the generated top-k subgraphs using the Python code provided below.

Program 8: Visualizing the Results

```
from PAMI.extras.visualize import graphs as vis
objVis = vis.graphDatabase(iFile='frequentTopkSubgraphs.txt')
objVis.plot()
```

16.9 Conclusion

In this chapter, we explored graph transactional databases and how they help represent and analyze complex relationships in data. We examined the difference between certain and uncertain graph databases, showing how they capture definite and probabilistic connections. We also covered key concepts like storage formats, conversion techniques, and tools for generating, visualizing, and analyzing graph data. A primary focus was discovering frequent subgraph patterns, which play a significant role in spotting recurring structures useful for pattern recognition, anomaly detection, and knowledge discovery. With applications in fields like social networks, bioinformatics, and chemistry, graph transactional databases are a powerful tool for making sense of interconnected data, opening doors for further research and innovation. The implementation code accompanying the examples in this chapter can be accessed on our GitHub repository: https://github.com/UdayLab/Hands-on-Pattern-Mining/blob/main/chapter16.ipynb.

References

1. Yan, X., and Han, J. (2002). gSpan: Graph-based substructure pattern mining. In Proceedings—2002 IEEE International Conference on Data Mining, ICDM 2002 (pp. 721-724).
2. Zhaonian Zou, Jianzhong Li, Hong Gao, and Shuo Zhang. 2009. Frequent subgraph pattern mining on uncertain graph data. In Proceedings of the 18th ACM conference on Information and knowledge management (CIKM '09). Association for Computing Machinery, New York, NY, USA, 583–592.
3. T. K. Saha and M. A. Hasan. FS3: A sampling based method for top-k frequent subgraph mining. 2014 IEEE International Conference on Big Data (Big Data), Washington, DC, USA, 2014, pp. 72–79.

Part III
Applications

Chapter 17
Discovering Air Pollution Patterns Through the KDD Process

Abstract Air pollution is a critical global environmental challenge, causing significant risks to public health, and ecosystems contributing to climate change. Extracting actionable insights from real-world pollution data is crucial for understanding and mitigating these risks. This chapter presents a detailed methodology for uncovering valuable patterns in air pollution data by applying the Knowledge Discovery in Databases (KDD) process. Using over five years of hourly $PM_{2.5}$ data collected from air quality sensors across Japan, we show how to preprocess, transform, and analyze this data using a combination of Python libraries such as Pandas, Scikit-learn, and PAMI. We walk through data acquisition, pattern discovery, and visualization, emphasizing how spatial patterns of high pollution areas can facilitate location-specific policy decisions. The findings highlight the effectiveness of combining data science techniques with environmental data to address global challenges. This chapter provides a replicable framework for applying the KDD process in various large-scale datasets, demonstrating its relevance to environmental monitoring and public health research.

17.1 Introduction

The previous sections of this book explored various types of real-world databases and the methods used to extract valuable patterns based on user interests. In Part 3, we focus on integrating the PAMI (PAttern MIning) [1] library with popular Python libraries, including Scikit-learn [3], TensorFlow [4], and Keras, to enhance knowledge discovery in large datasets.

One of the most pressing environmental challenges today is air pollution, which poses significant threats to human health, ecosystems, and the climate. Numerous organizations have deployed extensive networks of air quality sensors to tackle pollution, generating vast amounts of data. These datasets contain invaluable insights that could guide policymakers and environmental scientists in making data-driven decisions. However, extracting meaningful information from these datasets can be challenging due to the noisy and complex nature of real-world data.

U. K. Rage, *Hands-on Pattern Mining*,
https://doi.org/10.1007/978-981-96-6791-8_17

This chapter describes the Knowledge Discovery in Databases (KDD) process, a systematic method for extracting valuable knowledge from large datasets. Specifically, we focus on discovering pollution patterns in the $PM_{2.5}$ data collected over five years from various air quality sensors across Japan [2]. Figure 17.1 visually represents the KDD process, outlining the significant steps in identifying pollution hotspots and extracting actionable insights.

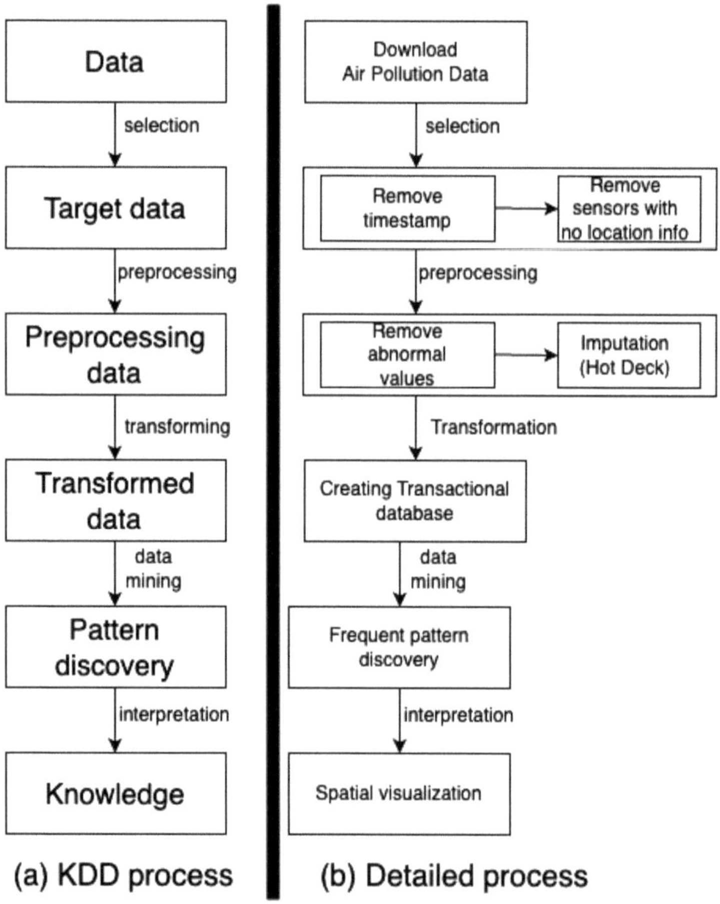

Fig. 17.1 The KDD process for discovering $PM_{2.5}$ pollution patterns in the data

17.2 A Step-by-Step Guide to the KDD Process

KDD involves several key steps, each critical to extracting valuable insights from raw data. Below is a step-by-step breakdown of how we apply KDD to air pollution data.

17.2.1 Step 1: Requirements

In this first step, you must install the necessary Python libraries to facilitate data processing, analysis, and pattern discovery. The key libraries we will use are:

- **Pandas**: Used for data manipulation and cleaning. This library will help us store the pollution data, perform basic analysis, remove unnecessary columns, and handle missing data.
- **Scikit-learn**: This library will be used for data preprocessing tasks such as imputation (filling in missing values) and other machine learning-related operations.
- **PAMI**: The core library for pattern mining. PAMI will help transform the dataset into a transactional database, mine frequent patterns, and visualize spatial patterns in the data.

To install these libraries, you can use the following command:

```
$ pip install pami scikit-learn pandas
```

The experiment uses over five years of hourly $PM_{2.5}$ data collected from sensors deployed by the *Atmospheric Environmental Regional Observation System (AEROS)*. This dataset covers the entire country of Japan. You can download the dataset from the following link: https://www.dropbox.com/s/wa8d1sujzlx56hh/ETL_DATA_new.csv.

17.2.2 Step 2: Selecting the Target Data

Once the data is downloaded, we load it into a Pandas Dataframe. The dataset includes a column labeled `timestamp`, which records the hourly intervals at which the data was collected. Additionally, each air quality sensor is identified by its unique location, represented as Point(X, Y), where X and Y are the geographical coordinates of the sensor.

Here is how you can load the data into a dataframe:

```
>>> import pandas as pd
>>> dataset=pd.read_csv('ETL_DATA_new.csv',index_col=0)
>>> dataset
```

The sample data stored in the dataframe is shown in Fig. 17.2. To clean and refine the dataset, we first remove the `timestamp` column and any attributes that do not

	TimeStamp	Point(139.0794379 36.3727776)	Point(139.1051411 36.3963822)	Point(139.0960211 36.4047323)	Point(139.0428727 36.3816035)	Point(138.9955116 36.33801589999999)	Point(139.342672 36.4105658)	Point(139.3526243 36.3695416)	Point(139.1945766 36.31351160000001)	Point(139.2076974 36.3034767)	...
0	2018-01-01 01:00:00	NaN	NaN	5.0	13.0	18.0	20.0	NaN	NaN	NaN	...
1	2018-01-01 02:00:00	NaN	NaN	11.0	12.0	22.0	15.0	NaN	NaN	NaN	...
2	2018-01-01 03:00:00	NaN	NaN	7.0	12.0	19.0	16.0	NaN	NaN	NaN	...
3	2018-01-01 04:00:00	NaN	NaN	5.0	11.0	16.0	11.0	NaN	NaN	NaN	...
4	2018-01-01 05:00:00	NaN	NaN	6.0	11.0	10.0	8.0	NaN	NaN	NaN	...
...	...	...	...	...	...	...	...	...	...	...	...
46000	2023-04-25 18:00:00	NaN	NaN	NaN	22.0	3.0	15.0	NaN	NaN	NaN	...
46001	2023-04-25 19:00:00	NaN	NaN	NaN	21.0	2.0	19.0	NaN	NaN	NaN	...
46002	2023-04-25 20:00:00	NaN	NaN	NaN	20.0	10.0	19.0	NaN	NaN	NaN	...
46003	2023-04-25 21:00:00	NaN	NaN	NaN	19.0	2.0	15.0	NaN	NaN	NaN	...
46004	2023-04-25 22:00:00	NaN	NaN	NaN	19.0	1.0	17.0	NaN	NaN	NaN	...

Fig. 17.2 Air pollution dataset

contain location information (i.e., columns with "Unnamed" in their name). This step ensures that we are only working with relevant data.

```
>>> dataset.drop('TimeStamp', inplace=True, axis=1)
>>> sensors=[col for col in dataset if 'Unnamed' in col]
>>> dataset.drop(columns=sensors, inplace=True, axis=1)
>>> dataset
```

17.2.3 Step 3: Preprocessing

In this step, we perform several data cleaning and normalization tasks to prepare the data for analysis. Specifically, we:

- Replace invalid or missing values with `NaN` to mark them for imputation.
- Remove sensors (columns) with more than 80% missing data.
- Use Hot-Deck imputation to fill in the remaining missing values by replacing them with values from similar records.

The following Python code snippet shows how we implement these steps:

```
>>> dataset.replace(['None', 'Nan'],np.nan,inplace=True)
>>> dataset.where(dataset <= 250, np.nan, inplace=True)
>>> dataset.where(dataset > 0, 0, inplace=True)
>>>
>>> threshold = 0.8 * len(dataset)
>>> dataset = dataset.dropna(thresh=threshold, axis=1)
>>> dataset = hotDeckImputation(dataset)
```

The function for Hot-Deck imputation is defined as follows:

Program 1: Hot-Deck Imputation

```
from sklearn.utils import shuffle
def hotDeckImputation(df):
    df_imputed = df.copy()
    for column in df_imputed.columns:
        missing_idx = df_imputed[column].isnull()
        non_missing_values = df_imputed.loc[~missing_idx,
        ↪ column]
        donor_pool = shuffle(non_missing_values,
        ↪ random_state=42).reset_index(drop=True)
        donor_pool_expanded = np.resize(donor_pool.values,
        ↪ missing_idx.sum())
        df_imputed.loc[missing_idx, column] =
        ↪ donor_pool_expanded
    return df_imputed
```

17.2.4 Step 4: Data Transformation

At this stage, we convert the dataset into a transactional database format for frequent pattern mining. $PM_{2.5}$ values greater than or equal to 35 are considered hazardous. Therefore, we set this as the threshold and convert the dataset into a binary format, where timestamps and locations with hazardous pollution levels are represented.

```
>>> from PAMI.extras.convert import denseDF2DB as db
>>> obj = db.denseDF2DB(dataset)
>>> obj.convert2TransactionalDatabase('TDB.csv','>=',35)
```

17.2.5 Step 5: Pattern Discovery

Now that we have transformed the data into a transactional format, we apply frequent pattern mining algorithms such as FP-growth to identify pollution hotspots. These hotspots represent locations where people were frequently exposed to high $PM_{2.5}$ levels.

```
>>> from PAMI.frequentPattern.basic import FPGrowth as ab
>>> obj = ab.FPGrowth('TDB.csv', 500)
```

```
>>> obj.mine()
>>> obj.printResults()
>>> obj.save('FPs.txt')
```

17.2.6 Step 6: Visualization of Patterns

Once we have discovered the frequent patterns, we visualize their spatial distribution. This helps us understand the geographical areas with the most consistent high pollution levels.

The following code generates a visualization of the pollution patterns:

```
>>> from PAMI.extras.graph import visualizePatterns as fig
>>> obj = fig.visualizePatterns('FPs.txt', 10)
>>> obj.visualize(width=1000, height=900)
```

Figure 17.3 shows the distribution of high pollution levels across Japan, highlighting both sporadic pollution events and consistent hotspots. Areas 1 and 2 in the figure represent regions with high pollution levels, suggesting that people living near these sensors are often exposed to harmful air quality. This repeated exposure in specific locations poses significant health risks to the local population.

On the other hand, Areas 3 and 4 show high pollution levels near individual sensors, but with a key difference: The sensors are far apart. While people near each sensor are exposed to harmful pollution, these areas are less likely to impact the same community, as the sensors are geographically distant.

These insights are valuable for guiding policy decisions. Areas 1 and 2, with concentrated high pollution, should be prioritized for interventions to reduce

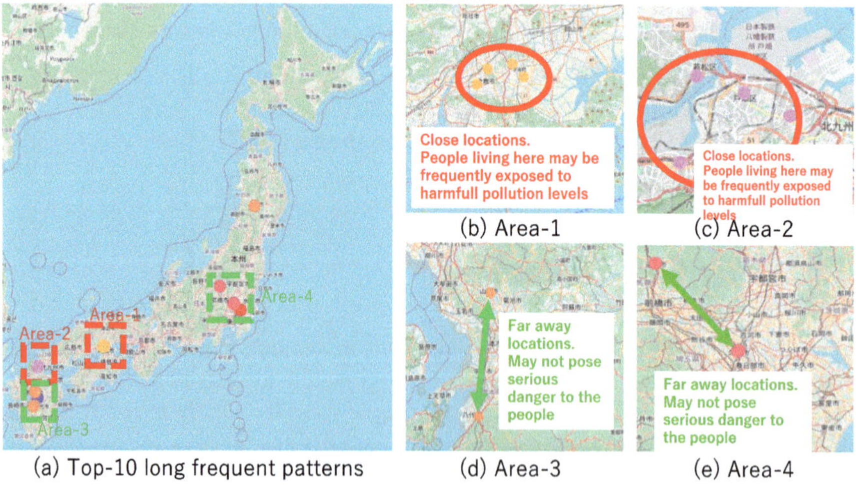

Fig. 17.3 Spatial visualization of the top-10 long frequent pollution patterns

exposure. In contrast, Areas 3 and 4, despite frequent pollution events, may require less urgent action, but further investigation is needed to understand the exposure risks fully. This information can help target pollution reduction efforts where they are most needed to protect public health.

17.3 Conclusion

In this chapter, we demonstrated how the KDD process can be applied to uncover pollution patterns in Japan using over five years of $PM_{2.5}$ data. By combining modern Python libraries such as Pandas, Scikit-learn, and PAMI, we successfully transformed raw pollution data into actionable insights through preprocessing, data transformation, frequent pattern mining, and spatial visualization.

The results revealed consistent pollution hotspots that can guide targeted policy interventions to reduce air pollution. This methodology underscores the power of integrating data science tools with environmental data to address pressing global challenges. Future work can expand this approach to other pollutants, apply predictive modeling techniques, and extend the framework to different regions, further enhancing its impact on global environmental monitoring and public health.

References

1. Uday Kiran Rage, Veena Pamalla, Masashi Toyoda, Masaru Kitsuregawa. PAMI. *https://github.com/UdayLab/PAMI*, 2024. [Online accessed 13-March-2025].
2. Ministry of the Environment. Atmospheric Environmental Regional Observation System: AEROS. *https://soramame.env.go.jp/download*, 2018. [Online accessed 13-March-2025].
3. Fabian Pedregosa, Gaël Varoquaux, Alexandre Gramfort, Vincent Michel, Bertrand Thirion, Olivier Grisel, Mathieu Blondel, Peter Prettenhofer, Ron Weiss, Vincent Dubourg, Jake Vanderplas, Alexandre Passos, David Cournapeau, Matthieu Brucher, Matthieu Perrot, and Édouard Duchesnay. 2011. Scikit-learn: Machine Learning in Python. J. Mach. Learn. Res. 12, null (2/1/2011), 2825–2830.
4. Martín Abadi, Ashish Agarwal, Paul Barham, Eugene Brevdo, Zhifeng Chen, Craig Citro, Greg S. Corrado, Andy Davis, Jeffrey Dean, Matthieu Devin, Sanjay Ghemawat, Ian Goodfellow, Andrew Harp, Geoffrey Irving, Michael Isard, Rafal Jozefowicz, Yangqing Jia, Lukasz Kaiser, Manjunath Kudlur, Josh Levenberg, Dan Mane, Mike Schuster, Rajat Monga, Sherry Moore, Derek Murray, Chris Olah, Jonathon Shlens, Benoit Steiner, Ilya Sutskever, Kunal Talwar, Paul Tucker, Vincent Vanhoucke, Vijay Vasudevan, Fernanda Viégas, Oriol Vinyals, Pete Warden, Martin Wattenberg, Martin Wicke, Yuan Yu, and Xiaoqiang Zheng. TensorFlow: Large-scale machine learning on heterogeneous systems, 2015. Software available from tensorflow.org.

Chapter 18
Discovering Futuristic Pollution Patterns Using Forecasting and Pattern Mining

Abstract This chapter presents a framework for forecasting air pollution levels and uncovering hidden patterns using machine learning and frequent pattern mining. We analyze over five years of hourly $PM_{2.5}$ data from Japan's Atmospheric Environmental Regional Observation System (AEROS). The dataset undergoes preprocessing, including handling missing values and normalizing sensor data. A long short-term memory (LSTM) model predicts future pollution levels across sensors. The forecasted data is then transformed into a transactional database, where hazardous pollution levels are identified using a predefined threshold value. The FP-growth algorithm is applied to extract recurring pollution patterns, highlighting critical pollution hotspots. These insights help policymakers develop effective pollution mitigation strategies.

18.1 Introduction

Building on the previous chapter, which explored Python-based pattern discovery in air pollution time series data, this chapter integrates forecasting techniques. We develop a model that predicts pollution levels and extracts meaningful patterns from the forecasted data. Figure 18.1 illustrates the framework for discovering pollution patterns from predicted data. The Python code of exercise is accessible at https://colab.research.google.com/github/UdayLab/Hands-on-Pattern-Mining/blob/main/chapter18.ipynb.

18.2 Step-by-Step Guide to Discovering Future Pollution Patterns

18.2.1 Step 1: Install Required Libraries

In this first step, the necessary Python libraries must be installed to facilitate data processing, analysis, and pattern discovery. The key libraries we will use are:

U. K. Rage, *Hands-on Pattern Mining*,
https://doi.org/10.1007/978-981-96-6791-8_18

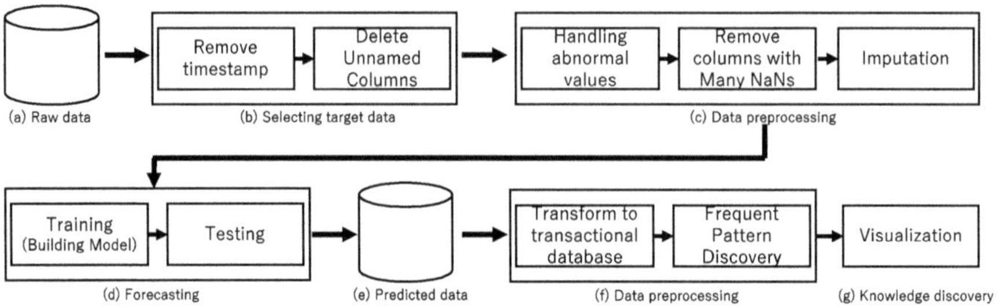

Fig. 18.1 The framework for identifying futuristic pollution patterns

- **Pandas** [2]: Used for data manipulation and cleaning. This library will help us store the pollution data, perform basic analysis, remove unnecessary columns, and handle missing data.
- **Scikit-learn** [4]: This library will be used for data preprocessing tasks such as imputation (filling in missing values) and other machine learning-related operations.
- **TensorFlow** [5]: This library will be used for building the prediction model for every sensor in the data. The long short-term memory (LSTM) algorithm is used for model building.
- **PAMI** [1]: The core library for pattern mining. PAMI will help transform the dataset into a transactional database, mine frequent patterns, and visualize spatial patterns in the data.

To install these libraries, you can use the following command:

```
$ pip install pami scikit-learn pandas tensorflow
```

The experiment uses more than five years of hourly $PM_{2.5}$ data collected from sensors deployed by the *Atmospheric Environmental Regional Observation System (AEROS)* [3]. This dataset covers the entire country of Japan. The readers can download the dataset from the following link: https://www.dropbox.com/s/wa8d1sujzlx56hh/ETL_DATA_new.csv.

18.2.2 Step 2: Selecting the Target Data

Once the data is downloaded, we load it into a Pandas DataFrame. The dataset includes a column labeled `timestamp`, which records the hourly intervals at which the data was collected. Additionally, each air quality sensor is identified by its unique location, represented as Point(X, Y), where X and Y are the geographical coordinates of the sensor.

	TimeStamp	Point(139.0794379 36.3727776)	Point(139.1051411 36.3963822)	Point(139.0960211 36.4047323)	Point(139.0428727 36.3816035)	Point(138.9955116 36.33801589999999)	Point(139.342672 36.4105658)	Point(139.3526243 36.3695416)	Point(139.1945766 36.31351160000001)	Point(139.2076974 36.3034767)	...
0	2018-01-01 01:00:00	NaN	NaN	5.0	13.0	18.0	20.0	NaN	NaN	NaN	...
1	2018-01-01 02:00:00	NaN	NaN	11.0	12.0	22.0	15.0	NaN	NaN	NaN	...
2	2018-01-01 03:00:00	NaN	NaN	7.0	12.0	19.0	16.0	NaN	NaN	NaN	...
3	2018-01-01 04:00:00	NaN	NaN	5.0	11.0	16.0	11.0	NaN	NaN	NaN	...
4	2018-01-01 05:00:00	NaN	NaN	6.0	11.0	10.0	8.0	NaN	NaN	NaN	...
...	...	...	...	...	...	...	...	...	...	...	...
46000	2023-04-25 18:00:00	NaN	NaN	NaN	22.0	3.0	15.0	NaN	NaN	NaN	...
46001	2023-04-25 19:00:00	NaN	NaN	NaN	21.0	2.0	19.0	NaN	NaN	NaN	...
46002	2023-04-25 20:00:00	NaN	NaN	NaN	20.0	10.0	19.0	NaN	NaN	NaN	...
46003	2023-04-25 21:00:00	NaN	NaN	NaN	19.0	2.0	15.0	NaN	NaN	NaN	...
46004	2023-04-25 22:00:00	NaN	NaN	NaN	19.0	1.0	17.0	NaN	NaN	NaN	...

Fig. 18.2 Sample air pollution dataset

Here is how you can load the data into a dataframe:

```
>>> import pandas as pd
>>> dataset=pd.read_csv('ETL_DATA_new.csv',index_col=0)
>>> dataset
```

The sample data stored in the dataframe is shown in Fig. 18.2. To clean and refine the dataset, we first remove the `timestamp` column and any attributes that do not contain location information (i.e., columns with "Unnamed" in their name). This step ensures that we are only working with relevant data.

```
>>> dataset.drop('TimeStamp', inplace =True , axis =1)
>>> unnamed_columns = [col for col in dataset.columns \
>>>     if 'Unnamed' in col]
>>> dataset.drop(unnamed_columns, axis=1, inplace=True)
>>> dataset
```

18.2.3 Step 3: Preprocessing

In this step, we perform several data cleaning and normalization tasks to prepare the data for analysis. Specifically, we:

- Replace invalid or missing values with `NaN` to mark them for imputation.
- Remove sensors (columns) with more than 80% missing data.
- Use Hot-Deck imputation to fill in the remaining missing values by replacing them with values from similar records.

The following Python program shows how we implement these three steps:

Program 1: Hot-Deck Imputation

```
import numpy as np
from sklearn.utils import shuffle

def hotDeckImputation(dataset):
    imputed_dataset = dataset.copy()
    for column in imputed_dataset.columns:
        missing_idx = imputed_dataset[column].isnull()
        non_missing_values = imputed_dataset.loc[missing_idx, column]
        donor_pool = shuffle(non_missing_values, random_state=42).reset_index(drop=True)
        donor_pool_expanded = np.resize(donor_pool.values, missing_idx.sum())
        imputed_dataset.loc[missing_idx, column] = donor_pool_expanded
    return imputed_dataset
```

```
>>> import numpy as np
>>> dataset.replace(['None', 'Nan'], np.nan,
>>>    inplace=True)
>>> dataset.where(dataset <= 250, np.nan, inplace=True)
>>> dataset.where(dataset > 0, 0, inplace=True)
>>> threshold = 0.8 * len(dataset)
>>> dataset = dataset.dropna(thresh=threshold, axis=1)
>>> dataset = hotDeckImputation(dataset)
>>> dataset
```

18.2.4 Step 4: Building Forecast Model

In this step, we apply the renowned LSTM technique to pollution forecast values of the sensors in the database.

Program 2: LSTM Model

```
import pandas as pd
import numpy as np
from sklearn.preprocessing import MinMaxScaler
import tensorflow as tf
```

```python
from tensorflow.keras.models import Sequential
from tensorflow.keras.layers import LSTM, Dense, Dropout, Input

# Assuming you have your 'dataset' loaded before this part
num_columns = int(input("How many columns would you like to predict? "))
columns_to_predict = dataset.columns[:num_columns]
all_predictions = {}

for index, column in enumerate(columns_to_predict, 1):
    print(f"Processing column {index}/{len(columns_to_predict)}: {column}")
    data = pd.to_numeric(dataset[column], errors='coerce').dropna().values.reshape(-1, 1)
    scaler = MinMaxScaler()
    data_scaled = scaler.fit_transform(data)
    X, y = [], []
    for i in range(len(data_scaled) - 1):
        X.append(data_scaled[i:i+1, 0])
        y.append(data_scaled[i+1, 0])
    X, y = np.array(X), np.array(y)
    X = X.reshape(X.shape[0], X.shape[1], 1)

    model = Sequential()
    model.add(Input(shape=(1, 1)))
    model.add(LSTM(units=50))
    model.add(Dropout(0.2))
    model.add(Dense(units=50))
    model.add(Dense(1))

    model.compile(loss='mean_squared_error', optimizer='adam')
    model.fit(X, y, epochs=10, batch_size=32, verbose=0)

    next_hours_input = X[-1:]
    next_hours_predictions = []

    for _ in range(24):
        next_hour_prediction = model.predict(next_hours_input, verbose=0)
        next_hours_predictions.append(next_hour_prediction.flatten()[0])
        next_hours_input = np.array([[next_hour_prediction.flatten()[0]]])
        next_hours_input = next_hours_input.reshape(1, 1, 1)

    next_hours_predictions = scaler.inverse_transform(np.array(next_hours_predictions).reshape(-1, 1))
    all_predictions[column] = next_hours_predictions.flatten()

predictions_df = pd.DataFrame(all_predictions)

output_path = 'LSTM_predicted_values.csv'
predictions_df.to_csv(output_path, index_label='Index')
print(f"Predictions saved")
```

18.2.5 Step 5: Converting the Predicted Multiple Timeseries Data into a Transactional Database

At this stage, we convert the dataset into a transactional database format for frequent pattern mining. $PM_{2.5}$ values greater than or equal to 8 are considered hazardous. Therefore, we set this as the threshold and convert the dataset into a binary format, where timestamps and locations with hazardous pollution levels are represented.

```
>>> from PAMI.extras. convert import denseDF2DB as db
>>> obj = db. denseDF2DB (predictions_df)
>>> obj. convert2TransactionalDatabase
>>>     ('TDB.csv','>=',8)
```

Next, we derive the statistical details of the transactional database to understand the distribution of items' frequencies. Understanding this distribution is crucial to specify an appropriate *minimum support* value.

```
>>> from PAMI.extras.dbStats import
>>>      TransactionalDatabase as tds
>>> obj = tds.TransactionalDatabase('TDB.csv')
>>> obj.run()
>>> obj.printStats()
>>> obj.plotGraphs()
```

18.2.6 Step 6: Pattern Discovery

Now that we have transformed the data into a transactional format and understood the distributions of its items and transactions, we apply frequent pattern mining algorithms such as FP-growth to identify pollution hotspots. These hotspots represent locations where people were frequently exposed to high $PM_{2.5}$ levels.

```
>>> from PAMI. frequentPattern .basic import FPGrowth
>>>      as ab
>>> obj = ab. FPGrowth ('TDB.csv', 15)
>>> obj.mine ()
>>> obj. printResults ()
>>> obj.save('FPs.txt')
```

18.2.7 Step 6: Visualization of Patterns

Once we have discovered the frequent patterns, we can visualize their spatial distribution by using the following code:

```
>>> from PAMI.extras.graph import visualizePatterns
>>>     as fig
```

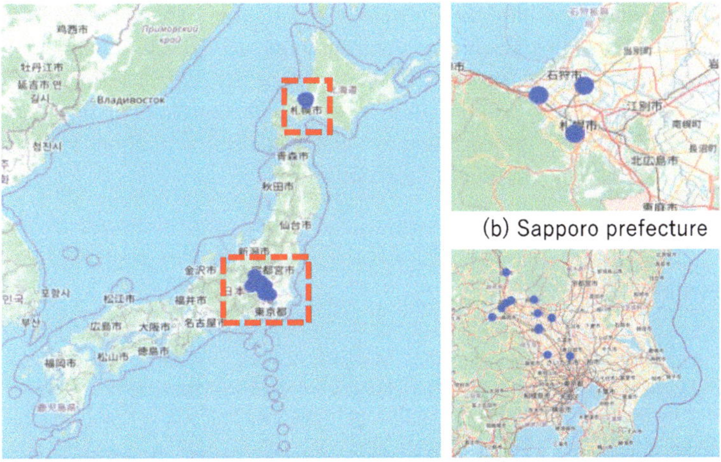

(a) Areas with probable high pollution levels
(b) Sapporo prefecture
(c) Saitama prefecture

Fig. 18.3 Spatially the areas that may witness high pollution levels shortly

```
>>> obj = fig. visualizePatterns ('FPs.txt ', 1)
>>> obj. visualize (width =1000 , height =900)
```

Figure 18.3 shows the distribution of high pollution levels across Japan, highlighting both sporadic pollution events and consistent hotspots. The two areas in this figure represent regions that may witness high pollution shortly. This repeated exposure in specific locations poses significant health risks to the local population.

These insights are valuable for guiding policy decisions. Areas 1 and 2, with concentrated high pollution, should be prioritized for interventions to reduce exposure. In contrast, Areas 3 and 4, despite frequent pollution events, may require less urgent action, but further investigation is needed to understand the exposure risks fully. This information can help target pollution reduction efforts where they are most needed to protect public health.

18.3 Conclusion

This chapter presented a systematic approach to forecasting air pollution levels and uncovering hidden patterns in the predicted data. By leveraging historical $PM_{2.5}$ sensor data and employing an LSTM-based forecasting model, we demonstrated how future pollution trends can be predicted with high temporal granularity. The transformation of forecasted data into a transactional database enabled the application of frequent pattern mining techniques, revealing critical pollution hotspots across different geographical regions. The visualized patterns provided actionable insights, highlighting areas with consistently high pollution levels and identifying potential health risks for nearby populations. These findings can serve as a foundation for policymakers and environmental agencies to implement targeted interventions and improve air quality management strategies. Future work could extend this approach by incorporating additional meteorological factors, refining prediction models, and exploring adaptive pattern detection methods to enhance the accuracy and reliability of pollution forecasts.

References

1. Uday Kiran Rage, Veena Pamalla, Masashi Toyoda, Masaru Kitsuregawa. PAMI. *https://github.com/UdayLab/PAMI*, 2024. [Online accessed 13-March-2025].
2. Wes McKinney. Data Structures for Statistical Computing in Python. Proceedings of the 9th Python in Science Conference. pp. 56–61, 2010.
3. Ministry of the Environment. Atmospheric Environmental Regional Observation System: AEROS. *https://soramame.env.go.jp/download*, 2018. [Online accessed 13-March-2025].
4. Fabian Pedregosa, Gaël Varoquaux, Alexandre Gramfort, Vincent Michel, Bertrand Thirion, Olivier Grisel, Mathieu Blondel, Peter Prettenhofer, Ron Weiss, Vincent Dubourg, Jake Vanderplas, Alexandre Passos, David Cournapeau, Matthieu Brucher, Matthieu Perrot, and Édouard Duchesnay. 2011. Scikit-learn: Machine Learning in Python. J. Mach. Learn. Res. 12, (2/1/2011), 2825–2830.
5. Martín Abadi, Ashish Agarwal, Paul Barham, Eugene Brevdo, Zhifeng Chen, Craig Citro, Greg S. Corrado, Andy Davis, Jeffrey Dean, Matthieu Devin, Sanjay Ghemawat, Ian Goodfellow, Andrew Harp, Geoffrey Irving, Michael Isard, Rafal Jozefowicz, Yangqing Jia, Lukasz Kaiser, Manjunath Kudlur, Josh Levenberg, Dan Mane, Mike Schuster, Rajat Monga, Sherry Moore, Derek Murray, Chris Olah, Jonathon Shlens, Benoit Steiner, Ilya Sutskever, Kunal Talwar, Paul Tucker, Vincent Vanhoucke, Vijay Vasudevan, Fernanda Viégas, Oriol Vinyals, Pete Warden, Martin Wattenberg, Martin Wicke, Yuan Yu, and Xiaoqiang Zheng. TensorFlow: Large-scale machine learning on heterogeneous systems, 2015. Software available from tensorflow.org

MIX
Papier aus verantwortungsvollen Quellen
Paper from responsible sources
FSC® C105338

If you have any concerns about our products,
you can contact us on
ProductSafety@springernature.com

In case Publisher is established outside the EU,
the EU authorized representative is:
Springer Nature Customer Service Center GmbH
Europaplatz 3, 69115 Heidelberg, Germany

Printed by Libri Plureos GmbH
in Hamburg, Germany